T0260349

ENVIRONMENTAL
GIS

Applications to
Industrial
Facilities

MAPPING SCIENCES SERIES

John G. Lyon, Series Editor

Aerial Mapping: Methods and Applications
Edgar Falkner, U.S. Army Corps of Engineers

Environmental GIS: Applications to Industrial Facilities
William J. Douglas, Environmental Resources Management, Inc.

Practical Handbook for Wetland Identification and Delineation
John G. Lyon, Ohio State University

ENVIRONMENTAL
GIS

Applications to Industrial Facilities

William J. Douglas

CRC Press
Taylor & Francis Group
Boca Raton London New York

CRC Press is an imprint of the
Taylor & Francis Group, an **informa** business

Library of Congress Cataloging-in-Publication Data

Douglas, William J.
    Environmental GIS, application to industrial facilities / William J. Douglas
        p.   cm. -- (Mapping sciences series)
    Includes bibliographical references and index.
    ISBN 0-87371-991-3
    1. Industry--Environmental aspects--Data processing. 2. Environmental protection--Data
        pro cessing. 3. Geographic information systems. I. Title. II. Series.
    TD1994.4.D68   1995
    658.4′08--dc20   94-19085
                                                                                                           CIP

© 1995 by CRC Press, Inc.
Lewis Publishers is an imprint of CRC Press

No claim to original U.S. Government works
International Standard Book Number 0-87371-991-3
Library of Congress Card Number 94-19085

This book is dedicated to my wife, Joan, and our family,
Bill, Stephanie, and Elizabeth.

ACKNOWLEDGMENTS

This book draws upon information and experiences acquired by the author over a period of more than two decades and in a variety of professional settings, the last ten of which have been in the environmental engineering field. In preparing the manuscript the contributions, critiques, comments, and editing of several friends and colleagues is sincerely appreciated.

Mr. Izak Maitin of Environmental Resources Management, Inc. (ERM) has been a key contributor in implementing the *EDGE* concepts and bringing them to reality, he also developed the screen image graphics from *EDGE* and Stratifact used as figures in Chapter 5, and has provided a critique and edit to the book. Mr. David Nale, President of Aerial Data Reduction Associates, Inc., reviewed the manuscript and provided valuable assistance relating to aerial photogrammetry and digital orthophotography. Mr. John Haasbeek of ERM assisted in the early development of *EDGE*, and Mr. John Kraeck of Graphic Data Systems Corporation participated in developing the first version of *EDGE* and in providing advice in its evolution. Mr. Richard (Pat) Hughes, retired Vice President of Administration at Avtex Fibers, reviewed an early draft manuscript from the point of view of the industrial community. Mr. Alex Lawes and Mr. Paul Nordquist of ERM edited the manuscript. Mr. Jeff Gorham assisted in providing digitized graphics for Chapter 5, and Molly McCreadie did a final proofread of the galleys.

Mr. Kent Patterson, President and CEO of ERM, Inc., provided ERM support to the development of the manuscript. The figures in Chapter 1 are variations of similar presentations of environmental regulations growth that have been developed at and used by ERM. ERM editors, word processors, and graphic artists contributed to preparation of the manuscript and graphics. The figures in Chapter 5 describing sub-surface visualization were contributed by and are used with the permission of Dynamic Graphics, Inc., ConSolve, Inc., and Intergraph Corporation. The software implementations and system integration activities that have been discussed within this book were conducted with the assistance and support of Graphics Data Systems, Inc. and Digital Equipment Corporation.

FOREWORD

During the final decade of the 19th century, a technology explosion was in evidence with the arrivals of the internal combustion engine, automobile, radio, and motion picture camera. The light bulb, phonograph, telephone, and ticker tape were only 10 years old, and the vacuum tube and airplane were less than 5 years away. This golden age of invention fueled the industrial and the economic growth of the 20th century. It profoundly affected that century, which was to bring the cataclysm of global war, the polarization of civilization into major political blocs, and the threat of thermonuclear warfare. A more subtle impact resulting from these effects has been the pressure on the global environment, and is attributable to the economic and population growth made possible by technology and the manufacturing and processing industries that have evolved to drive the economic standard of living in the industrialized nations.

During the final decade of the 20th century, information technologies are bringing increasingly greater levels of capacity into use by engineers, scientists, and managers for dealing with the business and technical issues. Personal computers, word processing, spreadsheets, electronic and voice mail, and cellular phones have created productivity levels that were unthinkable a few decades ago. At the leading edge of this information wave is the proliferation of graphics and visual computing, which finds acceptance by all computer users who are taxed by the sheer volume of information that is now available, and which the information highway has produced.

Viewing these two periods in perspective, we see that the solutions of the first era have played a major role in creating the problems of the second. The technical solutions of the past have been point solutions, each having a ripple effect in creating major problems in the future, somewhat like a very small interest rate causing dramatic accumulation of wealth over a long time period. The impacts of these solutions on society did not become items of concern until their effects were fully in evidence, having produced the social and environmental movements of the late 20th century. The industrial manufacturing and process industries that fuel the economy are incurring significant costs in dealing with the extensive volume of regulations that have resulted from these movements. New service industries have emerged to provide professionals with expertise for addressing these issues efficiently.

This book describes the application of information technology to address environmental, safety, and health (ES&H) information management in an integrated manner. The approaches that are described focus on dealing with information from an organizational or corporate standpoint; this means that the needs are not specialized to the ES&H area, but are an inherent part of managing the organization. Using available information sources and integrating the flow of information is preferred to growing new systems and creating redundancy. The emphasis is on the intuitive and graphical interface to the information by relating spatially to the geography and the physical objects that are a part of the facilities being managed. The objectives here are to assist industrial organizations in dealing with their environmental, safety, and health information management needs in the context of the overall corporate information flow and to make the information more easily accessible.

Emerging Geographic Information System (GIS) technology is being applied with increasing frequency to manage industrial facilities. This tool will become a standard in the 21st century for dealing with information. Those organizations who adapt and take advantage of the power of GIS will find applications that transcend the traditional facility management role. The ES&H applications identified in this book represent one important area. Creative organizations will find many others that have not yet been considered.

We cannot predict what the 20th century technologies will impose on the 21st century problem-solvers, but we should learn from history that the great solutions of mankind have led to significant problems for future generations. Solutions must be developed with a broader perspective, consciously looking for impacts that they may produce in a global, life cycle context. Those who are concerned with effective information management solutions can also adopt a global perspective in addressing the problems that they face.

THE AUTHOR

William J. Douglas is a Senior Program Director with ERM Program Management Company, a member company of the ERM Group, based in Exton, PA. He is responsible for the strategic planning, integration, and application of information management technologies and systems to client engagements, and has applied GIS database and graphics techniques to a variety of environmental project activities. He has designed the Environmental Data Graphics from ERM (*EDGE*) prototype system for environmental facility management and has managed its implementation and applications.

Dr. Douglas received his Bachelor's Degree in Mechanical Engineering from the University of Detroit, and his M.S. and Ph.D. in Engineering Mechanics from the Pennsylvania State University. He has had a varied career in the defense, energy, mining, and environmental fields and has authored numerous papers, articles, and publications in these areas. He is a registered professional engineer in the Commonwealth of Pennsylvania.

TABLE OF CONTENTS

Chapter 1
Information Needs 1
   I. Plant Personnel 2
      A. Organizational Approaches 3
      B. Data Consistency 4
  II. Managers 5

Chapter 2
Facility Management Systems 7
   I. GIS and FM 7
  II. Overview 8
      A. Infrastructure Management 8
  III. Applications 9
      A. FM and GIS 9
      B. Applications in Government and Industry 10
         1. Government Management and Planning 10
         2. Environmental, Safety, and Health Applications 11
         3. Resource Planning 11
         4. Commercial 11

Chapter 3
Intelligent Drawings 13
   I. Accuracy 13
  II. Drawing Structure and Organization 14
  III. Drawings and Photos from Aerial Surveys 17
      A. Drawing Specifications 17
      B. Quality Assurance 18
      C. Digital Orthophotography 19
      D. Access to Drawings 20
  IV. Coordination Among Project Participants 20
      A. Three-Dimensional Elevation Models 21
   V. Converting Available Drawings 22
      A. CAD Drawings 22
      B. Hard Copy Drawings 23
      C. Schematic Drawings 25
  VI. Integration of Drawings 26
  VII. Importing Map Files from Public Sources 27

Chapter 4
Database Management 29
   I. Compliance 30
      A. Quantifying the Requirements 30

    B. Implementing Systems 31
    C. Regulations, Documents, and Electronic Mail 32
       1. Access to Regulations 33
       2. Other Documents 33
       3. Integration with Electronic Mail 34
  II. Waste Site Characterization 35
    A. Managing Information for Site Investigations 35
       1. Data Types 36
       2. Historical Data Sets 36
       3. Historical Map Files 37
       4. Chemical Laboratory Analyses 38
       5. Mobile Field Analytical Services 42
       6. Field Data 42
       7. Aerial and Land Survey Data 44
    B. Information Resource 45
 III. Site Visualization 45
    A. High- and Low-End Approaches 46
    B. Software Tools for Visualization 48
  IV. Site Remediation 48
    A. Engineering Feasibility Studies 49
    B. Remedial Design 51
    C. Construction 51

Chapter 5
Approach for an ES&H Facility Management System 53
  I. Information Products 53
    A. Information Product Types 54
    B. Visualization 54
  II. Prototype System 55
    A. System Structure 56
       1. Hardware and Software 56
       2. Graphical User Interface 57
    B. System Functions 58
       1. Facility-Level Information 58
       2. Drawing Manipulation 59
       3. Drawing Access 59
       4. Database Access 59
       5. Commercial Compliance Software Integration 60
       6. Spatial Data Modeling and Visualization 61
       7. Image Access 62
       8. Cost and Schedule Data and Graphics 63
       9. Changing Drawing Features through Logical Queries 63
      10. Site Remediation Design 64
    C. Integrating with Specialized Software 64

1. Requirements and Alternatives 65
2. Evaluation of Alternatives 66
III. Executive Information Systems 67
A. Approach 67
IV. Example Information Products 68
A. Visualization of Contamination Plumes Using
High-End Workstations 69
B. Other Visualization Systems 70
C. Environmental Facility Management Graphics 73
D. Executive Information System 77

Chapter 6
Systems Integration 79
I. Factors Affecting Integration 79
A. Initial Conditions 79
B. Information Technologies 79
C. Integrating Available Systems 81
D. Structured Evolution 82
E. Drivers 83
II. The Players 84
A. Organizational Integration 85
III. The Information 86
IV. The Data Model 87
A. Purpose 87
B. Elements of Database Structure 88
C. Emissions and Discharges 88
D. Site Investigation 91
E. Executive Information System for Site Remediation
and Prioritization 91
F. Information Products 94
V. Regulatory Information 95
A. Federal and State Regulations 95
B. Local and Special Regulations 96
C. Integration Issues 96
1. Prototypes and Pilot Projects 97

Chapter 7
System Management 101
I. Computer Hardware 101
A. PC and Mini-Computer Workstations 102
B. Mainframes and Mini-Computers 103
C. Memory and Disk Storage 104
D. Monitors and Displays 105

    E. Digitizing and Screen Interface......................... 105
    F. Plotters and Printers................................ 106
  II. Software... 106
    A. Object- and Layer-Based Systems/Raster-Based GIS
       Data Models...................................... 107
    B. Software Structure 107
    C. Selecting Software 108
  III. System Use 109
    A. Administration Functions 109
    B. Access to Data and Information Products............... 110
      1. View-Only Access 110
      2. Response Performance 110
      3. Alternatives for Viewing Graphics.................. 111
  IV. System Implementation................................ 112
    A. Traditional Approach............................... 113
    B. Structured Evolution 114
      1. Vision 115
      2. Responsibilities 115
      3. Schedule and Milestones 115
      4. Measures of Performance 116

References 117

Index 119

Chapter 1

INFORMATION NEEDS

This chapter describes the needs of managers and plant personnel with regard to information relating to the facilities for which they have responsibility. It discusses approaches for dealing with regulatory compliance that are used by most organizations. Corporate upper-level management's concerns with organizational image, green movement, stockholder attitudes, and the need to be environmentally proactive are also addressed.

Since the early 1970s, there has been a continuous growth in the regulatory reporting requirements imposed by federal regulatory agencies and by state agencies that have followed the federal lead. These regulations were imposed in response to public demand for restoring the environment and for reducing the quantity of waste generated by industrial organizations from manufacturing processes. At the same time, there has been a superimposed increase in the regulations relating to safety and health that have added to the burden on industry for dealing with these issues. Figure 1-1 illustrates the growth of federal regulations relating to environment, safety, and health in the U.S. since the end of World War II. This regulatory burden has become significant, and the dramatic growth is evident since 1970. Another view of this effect is shown in Figure 1-2, which illustrates the growth in the number of pages of federal environmental regulations since 1970. During the 5-year period 1976 to 1980, new regulations were being imposed at the rate of about 6 pages per day. This has grown to a current rate of more than 20 pages per day. These conditions can be translated into the costs of compliance, since technical, legal, and management staff must be assigned to address the compliance assurance requirements that stem from the regulations. Moreover, there are associated information resources needed to support the management and reporting process that must be implemented in each industrial organization that falls under the regulations.

International regulation requirements have followed U.S.'s lead in most instances; however, in Europe there is a different relationship between industry and government, one that is not adversarial and is, conversely, based on a strategy of cooperation between the parties. This philosophy is also found in other international settings where government support of industrial development is different from that of the U.S., and has been a factor in this nation's international competitiveness in the consumer products market. Information

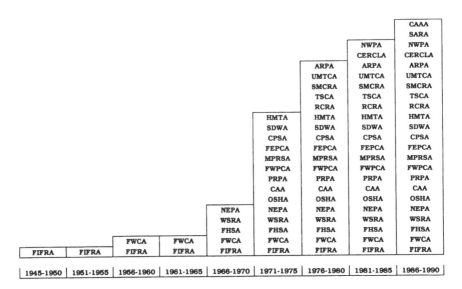

FIGURE 1-1. Federal regulatory burden faced by industry.

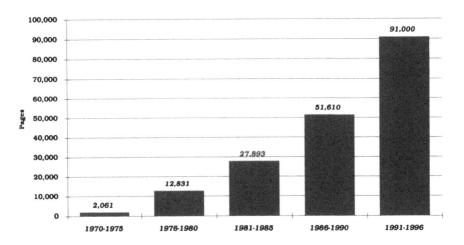

FIGURE 1-2. Growth in number of pages of federal environmental regulations.

requirements for regulatory compliance with waste management, safety, and health issues, however, will become increasingly important as these countries increase their living standards.

I. PLANT PERSONNEL

The regulations have imposed a variety of reporting requirements on industrial organizations and federal agencies such as the Departments of Defense and Energy, which have large industrial facilities. ES&H personnel at the plant level must interpret the regulations and the reporting requirements and collect information that can be used to comply with them. This requires quantitative measurement and summaries of items such as:

- Air emissions
- Wastewater discharges
- Soil and water contamination
- Solid waste generation
- Spills
- Hazardous chemicals use
- Chemical contamination levels in soils and water
- Worker exposure levels to toxic substances
- Worker and public right to know
- Accidents and worker risk
- Public risk from industrial activities

A. ORGANIZATIONAL APPROACHES

Because of the reporting requirements of the regulations and the tendency for an adversarial relationship with regulatory agencies, industrial organizations have adopted least-cost strategies for compliance. These have often been manifested by the application of limited resources to address these issues. Manual systems and simple spreadsheets are commonly used for developing the required computations and for preparing submissions to regulatory agencies.

As the environmental operations have grown in scope and size, there has been a trend toward integrating them with safety and health activities in many organizations to form an environmental, safety, and health department. Facility or plant managers often have an environmental group within the health and safety department, whose responsibility it is to deal with regulatory issues relating to the manufacturing processes that are being conducted at the plant. The sizes of these groups and their associated resources can vary significantly among organizations and even within the same organization. Decentralized approaches for dealing with these requirements have led to a myriad of procedures and systems. These include:

- Manual computation by plant personnel
- Development of applications using software tools by plant personnel
- Purchase of commercial software designed for reporting by plant personnel
- Corporate-developed systems for use by plants
- Licensing of application software by plants or corporate licenses for plant use
- Purchase of consulting services for assistance in producing regulatory reports

The strategy of compliance with regulations at minimum cost within a decentralized strategy has led to a great deal of diversity and inconsistency among systems. Organizations find themselves without a solid historic record of information, and regulatory agencies obtain data of varying quality with which to enforce regulations and with which to make future policy.

The ES&H managers at the plant, business unit, or corporate levels are also faced with significant challenges when it becomes necessary to address site characterization or remediation in their facilities. Consultants specializing in regulatory requirements associated with safety and health systems, risk analysis, ground water modeling, site investigations, or site cleanup are usually hired to provide technical and management support in addressing these problems. The work plans for these activities often require the collection and analysis of large quantities of data. For environmental site investigations, typical data requirements include:

- Site maps and aerial photos of waste areas
- Aerial and land survey results
- Soil and water quality data
- Stratigraphy
- Ground water elevations
- Geophysical logs
- Modeling results

This information must be managed to provide a solid basis for decision making so that the sites can be addressed efficiently and in a timely manner.

B. DATA CONSISTENCY

In many instances, a number of contractors may be hired to support an investigation; in other cases, one contractor may be involved in the early phases of a project and another may take over at a later phase. The environmental manager or a designated assistant may be charged with managing the activity. Furthermore, because of legal issues, the corporate legal departments must be involved in the activities. They, in turn, will sometimes hire lawyers who specialize in ES&H practice to manage the interaction with regulatory agencies. In some instances, the engineering consultants are directed by and interface with these legal firms regarding work on a site cleanup or the consequences of an unplanned discharge. Information managed in this manner takes on client privilege status.

These conditions provide an atmosphere for inefficiency that is manifested by the following results:

- Each party applies an independent and nonstandardized approach for collecting and storing data
- Contractors involved in subsequent phases of the project often dismiss earlier data sets as inconsistent and of poor quality
- Plant environmental managers are frustrated by the overwhelming quantities of data and the lack of real information
- The organization is left with volumes of paper and little ability to transfer the associated knowledge content to the next step in the process, or to future occurrences of similar work

In essence, the plant ES&H managers need spatially related information describing the relevant chemicals, processes, emission/discharge points, waste sites, and events that drive the management of compliance. They need to understand their operations and know what resources are being consumed by professional and clerical staff who produce the inputs and make the computations that are required by the regulations. They also need to provide corporate management with information to assist in corporate-wide management of environmental operations.

II. MANAGERS

A corporate manager having ES&H responsibilities for a number of facilities or plants is faced with the challenges of plant-level counterparts on a broader scale. The plants may be geographically dispersed so that problems cannot be resolved through direct observation, imposing an even more stringent need for high-quality and timely information.

There are yet other significant factors that affect the corporate environmental manager's information needs. The overall cost of dealing with regulations has risen from a nominal and nuisance consideration to levels that affect corporate performance and stockholder concern. Numbers approaching 4% of gross revenues are being experienced by corporations in their efforts to meet all of the relevant reporting and management requirements associated with ES&H issues, attracting the attention of corporate financial managers. Recognizing that their competitors are faced with the same problems, organizations have begun to explore approaches for achieving competitive advantages by dealing with them creatively.

As the visibility of ES&H costs escalates to the CEO and Board of Directors, another factor comes into play. Corporate top management is faced with image considerations in response to the growing "Green" movement. This affects how the organization will deal with concern for worker exposure to hazardous and toxic material, ozone layer creation, global warming, and public exposure to environmental threats from contamination of ground water, surface water, and from air dispersal of chemical emissions. Organizational policies are being developed to present a better image to the public in these areas. These policies relate to improved efficiencies in dealing with waste and also have a marketing component. This development is evident in advertising strategies, which have become increasingly friendly to the environment in recent years.

An item of concern to all managers is the occurrence of an unexpected incident at a corporate plant location. They face this possibility on a daily basis. Such an event has negative impact on corporate image and may result in extensive costs. Large oil tanker spills, rail and truck accidents, and toxic chemical discharges from plants produce headline stories leading to negative publicity and damaged corporate image. Corporate profits are affected by the disposition of these situations. Furthermore, they can have an immediate impact on the stock value, which brings with it the concerns of stockholders and the Board of Directors. It is essential that the likelihood of such problems be reduced to near zero, and that when a problem occurs, corporate management have high-quality information quickly to deal with regulatory agencies and the media under controlled conditions.

On the basis of these considerations, corporate ES&H managers are faced with information needs that can assist in:

• Formulating corporate image by making their products more acceptable to the public
• Making their plants more desirable work places
• Assuring stockholders and financial institutions that costs for regulatory compliance are being efficiently managed
• Dealing with unforeseen events where the organization could be exposed to public visibility

Addressing these needs is a continuing effort, and one that is receiving greater emphasis and attention. Approaches for meeting this need are the principal focus of the chapters that follow.

Chapter 2

FACILITY MANAGEMENT SYSTEMS

Facility management (FM) is an aspect of CAD and GIS technology that is growing in application. This chapter describes facility management applications and systems to set the stage for their application in an ES&H context. Applications to government and industrial organizations are described and examples of automated FM applications are given.

Since the advent of the industrial revolution, the efficient management of facilities has been the concern of manufacturing organizations worldwide. Industrial engineering and financial management methodologies have produced time and motion studies, process management, critical path scheduling, cost accounting, zero-based budgeting, and other techniques for dealing with improved efficiency and productivity. Recent developments in *total quality management* and *re-engineering* are all a part of industry's efforts to achieve market share and advantage and assure survival under the pressures of competition.

I. GIS AND FM

An automated mapping/facility management system (AM/FM) is a computerized integration of database management system (DBMS) technology with automated mapping (AM) to capture, store, retrieve, and display information graphically in order to facilitate its use and interpretation for reporting, planning, and decision making. A geographic information system (GIS) takes this concept further by providing tools for analyzing and relating the data spatially. GIS technology is rapidly growing because of the advances in computer hardware and software that make it available to a broad user base.

During the 1980s, the use of computerized graphical representation of industrial facilities became an emerging technology. The objective of the technology is to represent certain facility features in a manner useful to managers who are concerned with the acquisition and maintenance of the industrial components that are inherent to the manufacturing and production processes. These applications have been termed facility management (FM). By understanding these applications, it becomes evident that the same technology is applicable to ES&H issues, and these can be integrated within the context of the conventional FM applications.

II. OVERVIEW

An industrial facility consists of physical entities that are components of the processes of manufacturing, packaging, or distributing products. Industrial facilities are represented in plant drawings and processes, which illustrate their buildings, equipment, and infrastructure components. These can include:

- Pipelines
- Roads
- Fence/property lines
- Railroad lines
- Power lines
- Water/sewage lines
- Emergency systems
- Process diagrams

and other such representations that are essential for communicating ideas and displaying the information required to manage these operations.

A. INFRASTRUCTURE MANAGEMENT

The increased use of computer-aided design (CAD) drawings during the 1980s to represent facility entities gave rise to the term "facility management" (FM) as a specialized application of CAD drawings and DBMS technology. The term **infrastructure management** has recently become used to expand the concept to include larger and more general entities, such as:

- Highways and municipal roadways
- Bridges and tunnels
- Airports
- River and seaport facilities
- Public utility high-voltage lines
- Gas transmission and distribution lines
- Municipal water and sewer lines and treatment plants
- Electric and telephone distribution lines and stations

and other similar facilities that require planning, construction, and maintenance.

There are five principal information representations relating to the management of facilities and infrastructure:

- Maps, which illustrate the locations of these entities in relationship to geographic coordinates and geopolitical boundaries
- Drawings, which are generally more detailed representations of specific plants, sites, buildings, or processes
- Photographs, which can be aerial views at the map level or localized at the plant, site, or traditional photographs of equipment or construction features
- Data describing the engineering and financial attributes of the facilities in numerical or text format that can be manipulated to produce summaries and reports
- Documents containing policies, procedures, and specifications

Infrastructure management or FM systems are designed to store and provide access to these information types in support of planning and decision-making activities. The use of CAD drawings provides a means for locating the entities in a spatial context, which is how they are envisioned intuitively by those who work with them. Simple lists or text and numbers describing pumps in a plant or bridges along a highway are not adequate for effective management. The spatial aspect is essential to the process. Linking these lists to their spatial locations is the essence of FM. This same characteristic is true of ES&H information.

III. APPLICATIONS

A. FM AND GIS

FM, as an application of CAD drawings linked to database tables or attributes of CAD objects, has had its principal thrust in the state and municipal government areas in relation to planning and maintenance of land use data, property records, and public utilities. Recent applications have also addressed emergency responses to environmental incidents and spills, such as the Prince William Sound oil spill. Professional organizations have fostered the evolution of this technology. The Urban and Regional Information Systems Association (URISA) is a driving force that fosters information system development for local, regional, state/provincial, and federal government agencies. AM/FM International is focused on the application and advancement of the AM/FM technology primarily for energy and natural resource companies and government agencies. Communication and dissemination of information in this arena is achieved through professional conferences, publications, special programs, and local chapter organizations.

The variety of FM applications in this area is extensive and growing. This is somewhat of a renewed recognition of GIS. Earlier implementations in the 1980s met with some difficulties because of the under-powered systems and reliance on PC technology that was far below the capacities of the 1990s. Also, during the 1980s the emphasis was on the promise of the technology, and many of the data management issues were placed in the background, particularly the costs of installing and maintaining an effective geographic basis and integrated database.

B. APPLICATIONS IN GOVERNMENT AND INDUSTRY

Current FM applications can be found in any of the GIS journals and periodicals that are published regularly or by attending conferences that include technical papers on the subject. Examples of the types of activities that are being addressed through this technology are provided below. The reference journals that describe these and other applications are provided in the References.

1. GOVERNMENT MANAGEMENT AND PLANNING

- Municipal infrastructures — center-line drawings for streets, water, and sewer utilities linked to databases for integrated planning, construction, and maintenance management
- Regional planning — maps, land records, highways, redevelopment plans analyzed for regional impact
- Tax management — property maps, tax records, assessments for tax collection and planning
- Planning Olympic games — transportation planning and facility planning for athletes and spectators
- Emergency services — responding to fires, explosions, hazardous material spills, and other unpredictable events
- Decision support for transportation — traffic impact, signal optimization, and visual displays
- Organizational planning — restructuring and downsizing of government to evaluate alternative services and infrastructures
- Cost sharing — implementation of FM systems for municipal operations through integration of budgets from several participating agencies and organizations

2. ENVIRONMENTAL, SAFETY, AND HEALTH APPLICATIONS

- Environmental studies — evaluation of wetlands, erosion patterns, and watersheds using aerial digital orthophotography and satellite imaging
- Oil spill impact — remote sensing and surface-based evaluations of tanker spills, war disasters, and real-time management of emergency operations
- Wastewater management — integrated planning system including sewers, catch basins, ditches, and waterways for planning storm impacts
- Water quality planning — modeling soils, land use, and watershed characteristics to evaluate alternative scenarios
- Safety planning — mapping and locating unexploded shells on federal facilities using GPS and GIS tools
- Air emissions — modeling and display of dispersal and risk from air toxics on regions surrounding industrial facilities
- Process hazard analysis — linking drawings and databases to conduct Hazardous Operations (HAZOP) analysis for chemical operations required by OSHA

3. RESOURCE PLANNING

- Natural resource dynamics — food, fuel sufficiency, and watershed studies in areas of limited resources and conflict
- Forestry management — imaging and digital elevation modeling to evaluate damage to forests from the effects of fires, logging, pesticides, and acid rain, and to describe trends in forest resources
- Population planning — spatial distribution and mapping overpopulation and slums in underdeveloped countries using satellite imagery
- Habitat characterization — analysis of population and migration patterns to support preservation of endangered species
- Wildlife management — integrated land and species characteristics to evaluate wildlife and civilization interactions

4. COMMERCIAL

- Urban development planning — modeling diffusion of development processes to predict real estate growth patterns
- Market research — planning and estimating market patterns for services, products, and facility locations based on demographic, weather, and regional economic characteristics

- Siting and corridor planning — evaluating preferred locations for facilities or corridors for transportation or utility right-of-way based on population, land use, infrastructure, environmental impact, and facility operating characteristics
- Real estate — multimedia illustrations of maps and photographs to assist buyers

This is by no means an exhaustive set of applications. The fact is that so much information routinely used in government and industry has a spatial context; the applications are limitless. Computer databases have been used to manage these types of information for some time, and hard copy maps were prepared manually by cartographers or graphic artists to display it. With the evolution of GIS, functions can now be performed in the 1990s that were not possible in the 1970s and could only be achieved with costly equipment and less productivity in the 1980s. Continuing variety and increasing applications will provide even greater efficiencies and productivities from these systems.

Chapter 3

INTELLIGENT DRAWINGS

This chapter addresses the issues associated with creating and specifying drawings for ES&H FM applications, and for implementing the intelligence needed for linking them to databases.

The term *map* is commonly used for a representation of geographic facility features, while a *drawing* is a more general category of graphic that includes maps, site plans, building plans, and schematic representations. To achieve the desired functionality in an ES&H system, it is necessary to understand the relevant issues and interactions between drawings and data. These interactions are affected by both accuracy and logical content, which comprise the drawing's intelligence.

I. ACCURACY

A county or township map, facility drawing, pipeline diagram, or building plan can be created at differing levels of accuracy. The appropriate accuracy is dependent on the required use of the graphic representation.

When we ride the subway or try to find a gate in the airport, schematic diagrams are available to assist in navigating through the facility. These diagrams are simplified representations to assist in visualizing where we are and how to get to another location. On the other hand, when drawings are to be used for construction of a building, airport, or subway, it is necessary that they have a specified level of mathematical accuracy. The relative costs of constructing these drawings can vary by orders of magnitude, and it would be wasteful to use them inappropriately. Both schematic diagrams and the engineering drawings have applications in managing the facility.

A great deal of confusion exists with regard to the use of CAD drawings in supporting ES&H information management functions. If such drawings do not exist for a facility, it is believed that they will be too costly to create. If they exist in hard copy, it is felt that digitizing them will likewise be too time consuming and costly. It is important to recognize that for such applications the schematic diagram level of accuracy may be adequate and that technologies exist for creating drawing images from hard copy that may likewise be adequate.

There are advantages to using existing drawing files. If a digitized representation of the desired facility is available and intelligence has been added to the drawing, this can be of use to the ES&H manager. The disadvantages are that the drawing may be outdated or structured and organized in a manner that is inappropriate. As a consequence, it may take some effort to provide an effective relationship between the drawing and the database files that contain the relevant information.

II. DRAWING STRUCTURE AND ORGANIZATION

The term *intelligent drawing* is often applied to a graphical representation that is to be used in conjunction with a database to convey information. A map is a special type of drawing because of its coordinate system, which is a mathematical relationship between positions on the drawing that can facilitate spatial analysis of features or objects on the drawing. A drawing may contain a variety of information types, such as:

- Natural features
- Man-made features
- Modeled results
- Text and annotations describing components

These features are represented in *vector format* as points, lines, or chains (open sets of vectors constructed as nodes), and polygons containing areas (closed sets of vectors). This information can also be grouped or organized in different ways to facilitate its use. One standard approach is to partition the drawing into logical groups called *layers (phases)*. Typical layers are

- Buildings
- Roads
- Hydrology
- Topography

and other such drawing/map features. These are usually grouped so that any of these layers can be included or excluded to facilitate the intelligibility of the drawing, to aid in programming and automation, and to focus on those elements that are useful to the management objective. Moreover, each drawing layer can be organized further to facilitate the FM objectives. A layer can be a single entity or it can be created from more basic entities. Some systems use graphical representations called *objects* or even *facets* of objects. Objects may be grouped conceptually using a faceted naming convention. In addition, the drawing may have a topologic intelligence that defines logical relationships between points, lines, and areas such as connectivity and adjacency. These properties are important in GIS applications, but are not a part of CAD applications.

Drawing intelligence is imparted by the logical organization of layers and objects with the objectives of associating these entities with database tables and of effectively manipulating them in a mathematical and spatial sense. Figure 3-1 illustrates the concept of an intelligent drawing. Several electronic linkages are illustrated between drawing objects and other drawings and with data tables. These include:

- Each of the drawing objects has coordinates that provide a means for locating it spatially through the numbers that define its coordinate values
- The facility database table is linked to the facility drawing
- Sites on the facility are linked to site data — site 5 on the facility drawing has an associated site database table
- The facility drawing buildings are linked to floor plans — BLDG3 on the facility drawing has an associated floor plan drawing
- The floor plans define process blocks that are linked to process diagrams and chemical inventory tables — process block P-3a is linked to a full process diagram
- Process diagram emission points are linked to air and water emission database tables — emission point AE1 and waste water discharge point WD1 have associated database tables describing these emissions/discharges
- Floor plans are linked to chemicals that are stored in rooms on the plan — room RM 3-2 is associated with a chemical inventory table
- Waste sites have sampling locations which are linked to sample result database tables — sampling location B-2 has associated data records describing the analytical results for these samples
- Sampling locations are linked to ground water elevation tables — monitoring wells M-3 and M-5 are linked to their ground water elevation history tables
- Facility sites are linked to photographic images — WS3 has a photographic image with which it is associated

An FM/GIS (this term is used throughout the remainder of the book to describe the environmental facility management approach that is being advocated) system that is based on objects (object-oriented) provides a powerful means for interacting with data and graphics. Because of these linkages, it is possible for users to navigate easily and intuitively among objects by pointing at them and to retrieve associated data sets and locate information in a natural manner. It is best to impart at least a basic level of intelligence within the drawing during the formative period of the system design. It can be difficult to retrofit this intelligence once the system takes shape and becomes data intensive. It is not too different from human intelligence in this regard. Understanding this at the beginning of the data modeling activity is an important consideration in developing an effective FM/GIS system.

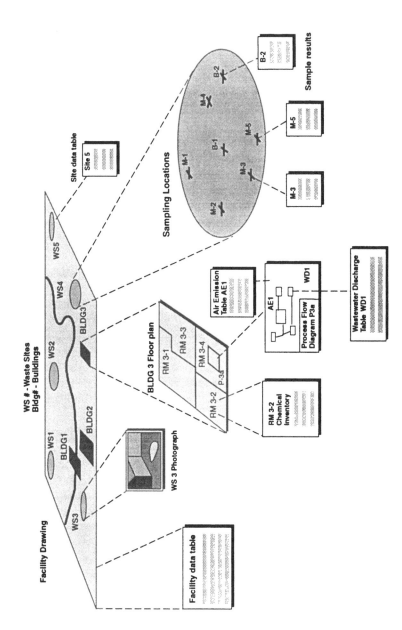

FIGURE 3-1. Intelligent drawings for environmental FM/GIS

An alternative graphical representation of a drawing or map is *raster format*, in which the drawing is simply a set of dots or points (in technical terms, a *bit map* representation), each having only an image variable such as shade intensity or color. Raster format drawings do not include intelligence, but they are useful in representing a continuous layer image of the facility in a single layer. Some systems allow for overlaying raster images on vector drawings or for layering raster images. To achieve this capability effectively, it is necessary to assure that control points are established in each drawing so that common features coincide, and that the mathematical projections of both images are consistent. This is the process of registration or registering the drawing. Differences in projection or errors in control points will produce errors in the overlays. For visualization of information, small errors may be acceptable for some applications.

The process of creating objects imposes some discipline on the CAD draftsman, which requires more effort than a simpler, unstructured approach. Some GIS systems are inherently layer-based and have difficulty representing complex objects. These systems are much less effective for use in FM applications and are likely to impose more difficulty than is desirable in addressing such applications.

Once the objects are defined, they must be named. An appropriate naming convention that relates to the database is the essential aspect of developing an intelligent drawing that can be effective in achieving the FM objectives of the system. The goal of relating graphical representations of facilities with comprehensive database tables requires the discipline of consistent structure, which may not be inherent in drawings that are made available by a corporate engineering department or by a previous aerial surveyor or engineering contractor. The imposition of standards for creating drawings is essential to effective FM applications.

III. DRAWINGS AND PHOTOS FROM AERIAL SURVEYS

When a site investigation is being conducted on an industrial facility, there is a need for a consistent base map that describes the property's man-made and natural features, referred to as *planimetrics*. These are usually obtained from aerial photos converted into drawings through the process of *photogrammetry*. Aerial surveyors who specialize in these procedures can be contracted to produce the base map, which is derived from aerial photographs of the site.

A. DRAWING SPECIFICATIONS

The base map specifications for a site must be stated clearly so that the appropriate levels of positional accuracy and level of detail are provided. The objective is to obtain map files that can be used throughout the site investigation and for the remedial design and actual site cleanup. This may involve substantive earth moving and topographical modifications. Drawing files used during the site investigation may not be appropriate for the remedial design, and additional costs may be incurred as a consequence.

When a base map is prepared from an aerial survey, it is necessary to develop thorough specifications for the deliverable products. The elements that need to be specified include:

• Horizontal accuracy — the maximum error that can be tolerated in spatial (x,y) positions (+/– ft.)
• Vertical accuracy — the maximum error that can be tolerated in elevations (z) above mean sea level (MSL)
• Contour interval — the desired spacing between elevation contours to provide a model of the topography
• Digital elevation model (DEM) structure (x,y,z) and gridding interval (x,y spacing)
• Coordinate system — the representation of values in feet or meters from a stated origin, which can be specified by identifying a standard coordinate system (e.g., state plane or U.S. Geological Survey [USGS] quad ID) or a local system
• Layering requirements for separation of planimetric features
• Topological structure describing relationships among points, lines, and areas (polygons)
• Digital format for the files so that they can be imported readily into the FM/ GIS system where they will be used
• Media on which the files are to be delivered — diskette/size/ density or tape format
• Documentation requirements — file names/layer names/ record counts
• Hard copy requirements — drawings/maps that will be provided in conjunction with the digitized file
• Delivery location — name/address where hard copy drawings and files will be delivered

B. QUALITY ASSURANCE

When drawings are received, it is important to ensure that they contain the required information. Project schedules can be affected when aerial or land survey steps are delayed by weather or other conditions. This can result in a subsequent high-intensity effort at a later date when there are pressures to load the files quickly so that they can be applied to immediate needs. A *quality assurance (QA) procedure* is essential to be sure that the files have been prepared correctly and that their use in creating information products can be authorized. A number of steps are appropriate to such a procedure, for example:

- Checking layers and counting records in the files and comparing them with the documented values
- Creating hard copy drawings from the digital files and comparing them with the delivered hard copy products
- Preparing topographic contours from the DEM file and comparing the results with field-surveyed test profiles and existing two-dimensional contour layers
- Checking text blocks for spelling and identification code accuracy

The QA procedure should be built into the project plan, and it is important for project managers to recognize its importance. There is usually a great deal of pressure on the FM/GIS staff to use files as soon as they arrive and a great deal of frustration for the project manager and field teams when the files are not properly formatted and structured in conformance to specifications. The importance of sound specifications becomes evident at this point, as does the realization that long-term use of these critical information resources warrants a reasonable effort to evaluate their quality before full-scale production of information products is initiated.

C. DIGITAL ORTHOPHOTOGRAPHY

A recent mapping technology that can be used in conjunction with CAD drawings is called digital orthophotography. Because aerial photographs are taken of uneven terrain from moving aircraft, there are scale changes induced in the photograph. The same area will appear larger if it is at a higher elevation, and an oblique view of the ground or surface feature is obtained if the camera is tilted due to aircraft motion. These effects cause photographic distortion (a 1-square acre should be the same size anywhere on the photo, but would be distorted due to the above-stated effects). Conventional orthophotography is the process of removing terrain- and motion-induced errors mechanically to produce without distortion high quality images that can be used as overlays with maps at the appropriate mapping scales.

More recently, with the evolution of GIS applications, orthophotos have been taken a step further. A pixel-based version of an orthophoto is produced by computer processing using stereoscopic aerial photos, a digital terrain model, and analytical triangulation to create appropriate gray-scale pixels of the photo in the selected pixel size (e.g., 1, 2, 4, ... ft.). In digitized format, the orthophoto can be used as a high-quality, raster-based, electronic overlay to a scaled vector drawing of the same area. Each pixel is precisely associated with a coordinate, and measurement between pixels is easily achieved. This can provide an approach to adding extremely accurate surface features within a drawing at a much lower cost than the process of vectorizing these features.

The digital orthophoto is an unintelligent raster image to be used for understanding the site. It can be stored on a magnetic disk or on a CD-ROM for access by the FM/GIS. The cost of the digital orthophoto and the associated storage requirements depend on the resolution (pixel size). If applied appropriately, this technique can provide cost-effective approaches for building the important base map component for an industrial facility as part of the ES&H information management system. The more costly process of photogrammetric compilation can be used on those features that need to be vectorized. Commercial GIS software vendors have adapted their products to using digitized orthophotos to provide accurate site backdrops for use in site investigation, facility management, and planning activities.

D. ACCESS TO DRAWINGS

Finally, there is the issue of *controlled access* to the files throughout the project life. Drawing files are used for a wide variety of purposes to produce information products; their continuing accuracy and accessibility must be assured. The procedures for achieving this objective are:

- There should be one responsible party for updating and maintaining digital drawing files
- Information products produced from drawings should identify the drawing file version/date that was used
- A backup copy of the drawings should be stored on diskette, disk or tape and in a different location than the production file

IV. COORDINATION AMONG PROJECT PARTICIPANTS

In many instances, drawing files arrive on diskettes in an unmarked envelope with no documentation. Interpreting the files under these conditions takes additional time, and poorly documented files are usually symptomatic of errors and inconsistencies in the files. Moreover, there may be several parties involved in producing the files:

- An industrial organization requesting the files (project manager, geology/ engineering groups, information management group)
- Environmental contractors
- Aerial photogrammetric firm
- Land surveyor

The levels of expertise within these organizations can vary significantly with regard to producing base maps, particularly in preparing three-dimensional digital elevation models. There must be a clear responsibility for communicating the requirement in writing and for coordinating among the project participants to eliminate or reduce unnecessary work associated with correcting the files and loading them into the FM/GIS system. For example, the land surveyor and the photogrammetrist should be provided with the locations of monuments or control points that are available on the facility, and the photogrammetrist must also understand the accuracy requirements so that the flights are specified at an appropriate altitude.

A. THREE-DIMENSIONAL ELEVATION MODELS

Most photogrammetrists are capable of producing acceptable two-dimensional base maps, which include planimetric features and elevation contours. There is sometimes confusion regarding the creation of appropriate three-dimensional (x,y,z) files. The process of creating a three-dimensional file requires the creation of a grid to describe the elevations at regular intervals. A gridded model of the property is a set of regularly spaced x,y sets having associated z values. Gridding provides a means for using data in a manner that facilitates decisions regarding the site, such as:

- Three-dimensional visualization
- Cross sections through topography and stratigraphy
- Volumetric computations
- Automated design of cut and fill construction

The two-dimensional representations of topography consists of contour lines defined by a set of strings of x,y coordinate pairs at constant elevations. These files contain the necessary information for obtaining a three-dimensional representation, but they must be processed to obtain a gridded model and can result in reduced accuracy. It is better to use spot heights from a DEM produced by photogrammetric methods; however, when these are not available, the two-dimensional data can be used. This processing consists of using a gridding algorithm, which conducts the following steps:

- Request a user-specified horizontal interval (x,y) at which to estimate elevation (z) values
- Use a mathematical estimating technique to produce an elevation (z) value at each x,y location
- Create a gridded (x,y,z) file for the topography that is based on the original aerial photos and/or the two-dimensional topography files

The gridding algorithms may be simple triangulation techniques, which interpolate each value by placing it on a triangular plane defined by a nearby set of three x,y points. More sophisticated *geostatistical* methods or other multipoint statistical or heuristic techniques are available in the public literature and in commercial software. The results can differ significantly when the data are sparse, and they tend to converge as the data sets become more robust and (even more important) regular.

It is important to recognize that the results of the gridding approach are modeled estimates of elevation rather than actual measurements. This means that using the DEM values to produce contours of constant elevation would result in differences from the contours provided as layers with the original base map. This is because contours developed from the aerial survey are produced by photogrammetric measuring techniques, while the process of gridding is a mathematical estimation technique. Producing contours from the gridded data is another mathematical estimation procedure that results in some degree of error, the magnitude of which depends on the nature of the topography and the gridding method. As a consequence of this situation, it is best to use the DEM values when conducting three-dimensional investigations, and rely on the original contours for simple two-dimensional viewing.

V. CONVERTING AVAILABLE DRAWINGS

In some instances, it may be appropriate to circumvent the need for new base maps by applying lower-cost techniques for producing acceptable base maps for a facility. Existing drawings within an organization's engineering or facility management departments may be applicable to the ES&H information system needs. These drawings are likely to require some modifications, since they would have been produced without the specifications for structure and content that are to be included in the ES&H facility management system applications.

A. CAD DRAWINGS

When a project is being initiated, the question of available CAD drawings is raised. Often, the only question is that of the specific software system on which the drawings reside or whether they can be loaded into the ES&H system structure. Most vendors of GIS and CAD systems offer utilities to import drawings from other systems; however, it is important to understand the structure of the existing drawings and the effort that might be required to convert them (i.e., to install the required intelligence to achieve the full functionality of the ES&H system). A variety of modifications might be required, such as:

- Converting from one digital format to another
- Transforming coordinates and scaling
- Partitioning into layers or phases

- Defining or grouping objects within layers
- Adding new layers/objects or integrating layers from other files
- Adding appropriate object names
- Modifying line styles
- Adding or deleting text blocks

Some of these can be automated fully or partially through utilities that perform the conversions. Others may require a CAD operator to apply drawing functions to change the file contents. Obviously, the level of effort and associated costs will depend on the extent to which any of the above is required. *It is important to recognize that the accuracy standards built into the existing drawing cannot be improved and must be adequate to the need if the conversion alternative is to be considered.*

The conversion process should be approached with the same considerations as specifying requirements for creating new maps or drawing files. The responsible person should provide information and advice to the project manager concerning all of the key issues, namely:

- Existing file types, contents, sizes, dates, formats, and accuracy levels
- Elements to be added to produce the required drawing intelligence
- Process steps and sequence for conversion and modification
- Responsible parties and schedule
- Quality assurance procedures

B. HARD COPY DRAWINGS

In some instances, hard-copy engineering drawings are available for a facility, rather than digitized CAD files. Their use will require conversion steps that are different from those associated with importing CAD files. There are three methods for extracting the information from these drawings for use as a basis in the ES&H system, namely, to:

- Digitize their contents using a CAD system digitizing table
- Scan the drawings and produce a raster image drawing for viewing access as a background
- Use automated digitizing services to create a vector drawing in the required formats

The requirements for digitized drawings in FM/GIS applications often create two diverse perceptions in those who are contemplating their application. These are:

- They can be scanned and loaded without difficulty; or
- Digitizing them will be too costly

It is important to understand the realities that are associated with these perceptions. Any drawing can be scanned to produce a raster image. The information content of the image may be adequate in raster form; but if an intelligent vector drawing is required for the FM functionality, the scanning serves no immediate purpose. Conversion of the hard copy drawing to a vector format is what is required.

Digitizing a drawing from an existing hard copy product can be achieved through three methods:

- Digitizing table and hard-copy map
- On-screen digitizing from an image
- Automated digitizing

The first approach consists of using a hard copy (paper or mylar) on a digitizing table and digitizing the selected objects. This approach is time and labor intensive, but provides the CAD operator with control over layering, object selection, and assignment of names and ID codes to objects. In addition, paper distortion and aging can introduce errors, which may be acceptable depending on the applications. Furthermore, if the hard copy scale is too high, line widths can be 10 ft or more than 50 ft, causing obvious digitizing errors.

By creating a scanned image of the hard copy, an electronic representation of the drawing can be displayed on the CRT screen. Digitization of this image follows the same steps as dealing with the hard copy on a digitizing table, except that the distortions can be eliminated, and the drawing can be zoomed to produce greater control over cursor locations relative to lines than in dealing with the direct hard copy. This reduces digitizing errors, but not the time and effort to complete the digitization process.

Automated digitization is an alternative to manual digitization and can be effective under some conditions. It starts with the creation of a scanned electronic image and then follows with processing of the image (e.g., *pixel clouding*) to produce vectors from selected lines. The process may be fully automated or it may require operator intervention. The quality of the digitized drawing generally will depend on:

- The quality of the source image, which must be interpreted by an intelligent processor that traces lines
- The existence of intersecting lines and broken lines, which can pose problems; separation of layers prior to line tracing can reduce these problems
- The complexity of the drawing; algorithms for interpreting the lines and text may become confused by very complex drawings

There are a number of service organizations which are competing to reduce the costs of digitizing drawings. They use automated, semi-automated, manual, and combinations of these methods to provide clients with digitization services. It is advisable to solicit quotations from such service organizations, particularly for large- scale digitization efforts. Moreover, the same concern for sound specifications is warranted as for the process of creating new drawings. Demonstration of their services should be made in a pilot or test mode by using both simple and complex drawings. These can provide benchmarks for estimating the time and costs for completing production digitization, and for making effective decisions concerning the appropriate approach for digitizing existing drawings.

C. SCHEMATIC DRAWINGS

Not all drawings used in the ES&H application need to be engineering drawings developed from aerial and land surveys. There is a place for simple schematic drawings that are less expensive to create and may be appropriate for some applications. A map of the U.S., a piping diagram, or a roof drawing can take on a wide variety of styles and formats. Produced to national mapping standards or accuracy levels of a few inches, they may be very costly; as schematic representations, they can be very inexpensive.

A roof drawing to locate emission stacks might be obtained from an engineering drawing of the plant; an alternative is simply drawing a rectangle or other polygon that is generally representative of the building and then placing stack objects at approximate locations. If the roof drawing exists, it may be possible to import it into the FM/GIS system; if not, the alternative may be a schematic diagram, without regard for scale or detail. If this approach is selected, the drawing can be created quickly and easily by a CAD operator; however, it is still important to apply standards for creating such files. The drawing could be used to identify roofs and stacks having certain attributes or data contents, but not to measure distances.

Process-flow diagrams represent another type of figure that is appropriate to ES&H FM applications. Such diagrams are inherently schematic; i.e., they are not to scale, and they represent a logical grouping of physical objects rather than a spatial grouping; however, the objects themselves (pumps, valves, tanks, compressors, etc.) are spatially locatable in the plant. A floor plan illustrating the locations of process blocks or of equipment is yet another representation that is appropriate for FM applications. Adding the connecting piping provides further process description, which may add value to the FM application. Photographs of equipment (scanned images) can be used to provide additional information content.

Schematic diagrams are intended to provide a means for navigating around the facility. As such, they need to have the same sense of spatial reference that the subway diagram has for the tourist. They must contain the correct objects and the appropriate object names. The base map for the facility may be developed to engineering scale, but when one selects a specific building or process within the facility, the issue of engineering accuracy becomes negotiable. In other words, if the issue of distance and direction is important (as for modeling air emissions from a stack), then the roof diagram could be a schematic, while the base map showing the building location could be produced to engineering standards. If roof drawings to engineering standards can be imported, there would be no disadvantages to using them. The general guideline is to understand the objectives of the application and to select the least-cost approach for achieving the objectives.

VI. INTEGRATION OF DRAWINGS

It becomes evident that, as the ES&H application evolves, the use of available engineering drawings becomes a key issue of concern. Clearly, it is not advisable to have several drawing standards within the same organization or to have a cumbersome exchange of drawings among organizational departments. Engineering organizations typically deal with CAD drawings for design and construction purposes. The standards that have been imposed are based on these requirements and have an institutional basis and history of utilization. The need for these drawings in an ES&H context creates an organizational interaction that was previously not present. The involved parties will react in a manner that is consistent with their individual goals and functions within the organization. These can be defensive and territorial or positive and open. The goal is to achieve an organizational benefit without imposing unacceptable costs.

The development of drawings for the FM application requires a thorough understanding of the existing organizational entities and its capabilities for creating, storing, and making drawings available. This is best achieved by meeting with those who are responsible for maintaining corporate drawings and establishing a working relationship that facilitates the process of moving forward. By understanding what is needed and what is available, it is possible to create a situation whereby the current procedures are not replaced, but evolve to meet a new requirement.

In order to achieve this objective, it is necessary to address the issue of requirements and specifications. The ES&H application can have certain drawing and data structure requirements that can be met in different ways:

- By acceptance of existing engineering CAD drawings and conversion by the ES&H group

- By revision of drawings by the engineering group to meet ES&H need
- By modification of the ES&H system to adapt to the engineering CAD system and standards

In other words, it is essential to be flexible and to consider all of the alternatives before adopting a course of action. Each of these will have associated costs (file conversion, software modification, CAD operator time, historic drawing revisions, etc.). The goal is to achieve an organizationally optimal result, which may require compromise on each side. The decision may be made to operate two distinct systems, with little or no integration. This may be less expensive from the standpoint of each of the involved parties, but it might not be economical from a long-term corporate view. When this situation occurs, it is advisable to have a higher level of management involved that can provide a global view of the impact of the selected approach.

VII. IMPORTING MAP FILES FROM PUBLIC SOURCES

For some applications, commercial services are available to provide map files that can be used as drawings to support ES&H applications. Public agencies such as the U.S. Geological Survey or state surveys have taken steps to make map files available in digital format. EPA Comprehensive Environmental Response, Compensation and Liability Act (CERCLA) data relating to the location and nature of hazardous waste sites are also available from private commercial sources. Satellite photos and infrared images are commercially available at regular spatial and temporal intervals, and these services may also be commissioned for specific needs. With the continuing emphasis on geographical representation of data, these services continue to grow. Private sector organizations also offer value-added services based on publicly available government data sources.

It is important to recognize that publicly available map images or data sets must usually undergo some level of conversion before they can be useful within an FM system. First, the images are provided to a certain level of accuracy and in a specified projection. If they are to be used in a stand-alone mode, then they can simply be imported and viewed on standard imaging software. If they are to be integrated, then it becomes necessary to assure that the projections are appropriate. If a geo-based data set is acquired, often its locations are specified in latitude, longitude values, which must be converted to the desired projection if the locations are to be displayed on a map. The guidelines for acquiring and importing public data sets for use in FM/GIS applications are:

- Maintain awareness of what is available by subscribing to periodicals and attending FM/GIS conferences that provide information on such services
- Obtain an understanding of the specifications and format for the products that are available, and how they relate to the standards employed within your organization
- Acquire or develop the appropriate mathematical transformation software, which may be available within the FM/GIS to transform coordinates or to adapt images and data that are to be displayed on base maps and used in overlays
- Consider the information value of the product in impacting on your organization's decisions before making the acquisition
- Understand accuracy and reliability (names, locations, representations) issues when dealing with commercial products and the compromises associated with their use

Chapter 4

DATABASE MANAGEMENT

This chapter describes database management systems and commercial software that meet the needs of environmental managers relating to compliance management, including ad hoc reporting as well as report generators to produce actual submissions to federal and state regulators. It also describes database requirements for managing data on hazardous waste sites, including soil and water chemistry, mapping, geology, risk, geophysics, drilling logs, and laboratory interface. The application of FM/GIS and modeling techniques to site remediation analysis and design activities is also discussed.

Industrial ES&H organizations have addressed the requirements imposed by regulations through the application of computerized systems for collecting, storing, and reporting information for compliance with federal and state regulations. There has been a continuing evolution in this area over the past 20 years, fueled by the growth in regulations on the one hand and the rapid developments in information technology on the other. This evolution has taken place along three tracks:

- Compliance management systems focusing on data collection and reporting associated with specific regulations
- Access to regulations in electronic format to support ES&H management and planning and response to changes and emergencies
- Data management systems to store and apply data that are appropriate to site investigations and remediation design

Developments along these three tracks have produced a number of systems that focus on specific regulations. In some instances these have been integrated to provide a commonality that applies to several regulations. For the most part, however, there are a variety of software systems and approaches in place in most organizations, including commercially developed and in-house-developed systems. Driven by policies of continuing response to new regulations, organizations have often implemented a piecemeal, point solution approach to system development. This has led to a number of inefficiencies including duplication of data and incompatibility among systems.

I. COMPLIANCE

Compliance management systems for ES&H applications are driven by a number of regulatory requirements for reporting data at regular intervals. Typically, a regulatory requirement is defined for which information management needs becomes evident. Examples of these are:

- Superfund Amendment and Reauthorization Act (SARA), Resource Conservation and Recovery Act (RCRA), and Solid Waste Disposal Act (SWDA) requirements for underground storage tank management as specified under 40 CFR Parts 80 and 281
- PCB equipment inventorying under Toxic Substances Control Act (TSCA) § 6(e) as specified under 40 CFR Part 761
- Emissions reporting for air, water, and solid waste under 313 Form R requirements
- Chemical inventory reporting under SARA requirements and OSHA
- Storage of and access to material safety data sheets (MSDS) to support emergency operations and right to know regulations under SARA Title III
- Clean Air Act Amendments Title V requirements for emission inventories and fugitive emission reductions.

A. QUANTIFYING THE REQUIREMENTS

A comprehensive set of specific requirements such as these is applicable to each facility within an organization. The ES&H manager at the facility should have a description of these requirements and the organization's approach for dealing with them. A summary of requirements and associated procedures and approaches is often not available; rather, approaches evolve in the organization for dealing with these requirements, they are adopted, and they become fixed within the organization. There is usually little integration among groups dealing with different regulations and sometimes even among plants dealing with the same regulations.

A planning tool that is appropriate to the compliance management function is a simple list that can be stored in a database system or in a spreadsheet. Its contents would include:

- Law and regulation reference for the requirement
- Responsible parties for addressing the requirement
- Reports and reporting frequencies
- Effort, schedule, and procedures employed to submit reports
- Database tools and systems used
- Interrelationships among systems

More details could be included as they become available. This information can be valuable in providing management with visibility into the current status of activities and the resource requirements that are imposed on the organization. This is a first step in gaining corporate knowledge of the impact of regulatory requirements on compliance. Such information is not often available to ES&H managers explicitly, and they may have only a general or perceptual understanding of the overall costs of compliance and its relationship to organizational performance.

B. IMPLEMENTING SYSTEMS

Compliance management systems are somewhat like accounting systems in that they require collection and storage of data and preparation of reports. Reporting may be hard copy or electronic. Regulatory agencies are expanding their acceptance of electronic deliverables through diskette or telecommunications submittal as these technologies become routine means for being more productive. Even the IRS accepts electronic tax returns.

Implementing systems for compliance management can be achieved through:

- Design and development, starting with requirements definition and ending with beta-tested software
- Acquisition of commercial software
- An integration approach, which may involve both of the above steps

Design of new software is expensive and time-consuming. Moreover, MIS staff are typically not familiar with ES&H requirements and have to go through a significant learning process to specify an appropriate set of requirements that can be implemented. In many instances, MIS staff adopt strategies such as requesting demonstrations of commercial systems to gain understanding of requirements or developing preliminary requirements and then requesting proposals from software vendors or system developers. The time required to implement a new system can be a year, or even several years for a complex system. The advantages are that the system is customized to the organization's requirements and is integral with the hardware and software platforms that are supported by the MIS department.

Acquisition of existing software permits a faster implementation at a lower cost. Commercial software systems offer these advantages by responding to a regulatory need and spreading development costs across many users. The maintenance costs are also lower, since software vendors offer updates and telephone support as part of their licensing policies. However, there may be compromises required, such as adopting an unfamiliar software platform, acquiring new hardware, and implementing new procedures for managing information.

An integration approach starts with developing a firm set of requirements, which may be served in part by existing systems and procedures but requires enhancements to them. After developing the requirements, it is useful to obtain demonstrations of software and evaluate them against the requirements. Subsequently, an organization can implement pilot projects to test one or more of the preferred alternatives at a limited set of locations, say one or two. The pilot projects may include enhancements to the process, such as:

- Using existing systems in conjunction with commercial software
- Developing file exchange procedures to move data from existing data sets into commercial software
- Converting commercial software components to a company's preferred database management system
- Creating links between commercial database management software and in-house FM/GIS systems

Commercial software for environmental applications has been catalogued in the *Environmental Software Directory* published by Donley Technology.[18] This document provides a basis for evaluating software systems that are being considered for integration within an organization's environmental information management system (EIMS). The 400-page document describes more than 1000 software packages. Evaluating even a small portion of these is an exhaustive undertaking. A structured methodology should be applied when considering acquisition of software in this manner.

Evaluation criteria relating to the performance requirements and the organization's environmental business operations, software platforms, and hardware should be used to select software for evaluation. Vendors can then be asked to provide additional information including brochures, demonstration diskettes, and videotapes. Further evaluation would screen the vendors down to a selected set for which demonstrations would be requested. From these demonstrations, system components would be selected for installation to conduct pilot projects to test them under operating conditions. It is neither necessary nor prudent to commit to full implementation prior to this step. Pilot projects should have additional evaluation criteria to solicit comments from in-house users of the system. Full system roll-out would be authorized following successful attainment of expectations during the pilot project.

C. REGULATIONS, DOCUMENTS, AND ELECTRONIC MAIL

The regulations themselves provide an important information source that is inherent to the ES&H need. Regulations are available in hard copy and in electronic format and can be accessed via personal computer, workstation, or terminal to a mainframe. Software associated with such systems can be generic, i.e., applicable to any document, or it may be a customized application for ES&H needs.

1. ACCESS TO REGULATIONS

Regulations are simply documents having specific information that drives requirements for ES&H systems. They are used by corporate lawyers, ES&H managers, plant staff, and other professional staff involved in complying with requirements. As such, they may require a broader level of access than compliance or site investigation systems. These documents exist in word-processed format with their publishers, but the typical user is not interested in this format. ES&H document usage requires *read-only* access. Users are interested in reading and interpreting, not in changing these regulations (changes of this type would certainly not be enforceable!).

The regulations should be located on readable mass storage media; CD-ROM technology is best suited to this application. Because they are voluminous and extensive, it is necessary to have a means for navigating through the regulations, which is best achieved through an indexing system that links sections of the text to their numerical designations (e.g., CFR 40.281). In addition, it is desirable to search documents for key words and identify their locations in the text, providing a means of reviewing the document in a particular context.

There are a number of generic software systems available that are capable of providing access to regulations or other documents on magnetic or CD storage media. Service corporations also offer specific regulatory information in either full-text or abstracted format through annual licensing, which can include regular updates to assure currency in the continually changing regulatory climate.

2. OTHER DOCUMENTS

Document storage can be achieved either through text files, based on a word processor or text management database format, or through the use of imaging. Inbound material safety data sheets (MSDSs for chemicals that are manufactured by another organization) are an example of documents that are best stored as images, since their text or numeric fields should not be modified. Corporate ES&H policy documents and program plans, emergency procedures, training materials, and reference books are examples of documents for which electronic access can provide efficiency and productivity enhancements in large organizations.

Graphical documents, such as engineering drawings, represent another type of document that is applicable to the ES&H function. Storing such documents in an effective electronic document management system (EDMS) is appropriate. The proliferation of electronic documents is cluttering many personal computers, networks, and electronic mail systems, and many users are not up to the task of maintaining these files. The paper explosion has been superseded by an electronic explosion where disks containing multimegabytes have replaced file cabinets.

3. INTEGRATION WITH ELECTRONIC MAIL

Electronic mail (E-mail) is yet another component of document management that can have specific ES&H applications. E-mail is commonplace in today's office environment to move memoranda, announcements, and documents around an organization without paper transfer. E-mail can extend throughout a location using local area network (LAN) technology, which links groups of computer users to common files and other computer resources. It can also extend throughout a complete organization using wide area network (WAN) technology, which relies on telecommunications over leased or dial-up lines. New software tools are now available to integrate E-mail with other software, to perform specialized functions such as:

- Announcing changes in regulations based on updates to the regulatory document system or regulated chemicals recently published in the *Federal Register*
- Transmitting announcements of training programs to specific staff based on personnel records and new regulatory requirements for training
- Providing project labor and expense information from time sheets and accounting systems as well as laboratory analytical results from technical databases to project managers who are located in the field during site investigations
- Distributing test results for a site investigation or an air emission test from the field to the project database or to project personnel as soon as they are loaded into the project database

Such applications represent an integrated approach to information management. The concept of groupware has emerged as a network software technology for group interaction with electronic documents. Such documents can be distributed as text, data, or images, depending on the application. They can be provided in read-only format if this is appropriate. Groupware and E-mail are complementary technologies that can find important applications in the ES&H arena. By integrating these technologies, new and productive applications can be implemented. The same information can be distributed via hard copy mail or diskette, but to those who have become accustomed to dealing with E-mail, it is clearly a preferable medium of information exchange.

II. WASTE SITE CHARACTERIZATION

Many organizations are required to undertake hazardous waste site investigation under CERCLA and SARA regulations. These have resulted in the application of a variety of databases, spreadsheets, and modeling systems to support the investigation teams. A number of data types and related information products are used to support these investigations, such as:

- Soil and water chemical analysis results
- Drilling logs and stratigraphic data
- Ground water elevations
- Geophysical logs
- Topography and site planimetrics
- Historic aerial photographs
- Fence diagrams
- Concentration contours
- Time lines and trends for site variables

The management of such information has become a concern that has been addressed in varying degrees by organizations faced with hazardous waste site management.

A. MANAGING INFORMATION FOR SITE INVESTIGATIONS

For the most part, site investigations and subsequent remediation activities are conducted by environmental contractors who specialize in the interpretation of the regulations and in conducting such regulations in accordance with EPA and state requirements. Some organizations have engineering/construction departments that manage and participate in such activities in varying degrees. The environmental contractors have adopted a variety of approaches for handling data in support of these investigations. These have included:

- Database management systems developed and used by environmental contractors
- Commercial database management systems purchased and used by environmental contractors
- Spreadsheets and statistical software to store and manipulate data
- Specialized geological software packages
- CAD systems to prepare drawings describing site characteristics

Generally, the data management approach has consisted of applying several unconnected systems. Information exchange among these systems has usually not been planned in advance and has often been accomplished through diskettes containing ASCII files produced by one software module and intended for use in another. As such, the approach has been one of loosely connected modules rather than a truly integrated system. This can result in inconsistencies, delays, and inefficiencies in producing the appropriate information products from the available data.

1. DATA TYPES

There are seven broad categories of data (both tabular and spatial) that are applied in conducting site investigations:

- Historical data from prior investigations
- Historical maps/drawings available from prior investigations or from company engineering files
- Chemical laboratory analyses from current investigations
- Chemical analyses from on-site mobile units used to screen samples in the fields
- Field data not requiring laboratory processing, such as drilling logs, ground water elevations, etc.
- Land survey data locating sampling locations and other important site features
- Aerial survey data including photos, base maps, and three-dimensional digital elevation files

2. HISTORICAL DATA SETS

Site investigations ranging from preliminary assessments and inspections to in-depth remedial investigations have been conducted since the enactment of CERCLA to produce millions of documents describing site conditions where hazardous waste has been disposed. These activities have produced documents to comply with EPA requirements for conducting such investigations. The scientific information contained in the documents is usually available in hard copy and in a word processor file. There may be spreadsheet files to document tables or appendices included with the documents to support the technical findings. Database tables have become more common during the 1990s.

Historical data that may have been developed from the efforts of a prior contractor are usually difficult to use. The reasons for this are:

- The data may require considerable effort to reconstruct if not available in electronic format
- The data structure, even if available electronically, may be poor and incomplete
- Results may not be easily relatable to spatial coordinates
- Supporting quality assurance information may not be available
- The data may not be associated with identifiable locations on the site for which coordinates are known
- Contractors suspect the overall quality of historic data sets provided by other contractors and are not comfortable in using them
- Improved laboratory analytical methods and changes in analytical standards make these data sets different from those that are current

Historical data sets provide a point of departure for conducting subsequent site investigations. The value of including these in the electronic database for a current investigation depends on the relative costs incurred, the quality of the data, and the benefits that are expected from doing so. There are advantages to having all data accessible consistently for comparative purposes and for planning and identifying trends and changes in site conditions. Historic data should be entered into the organization's facility database. These records, however, should be identified by their source and date of collection so that they can be included or excluded from statistical analysis and summaries or displays in an appropriate manner.

3. HISTORICAL MAP FILES

Historical maps and aerial photos of facilities where site investigations are being conducted provide an important information source to the investigation team. They are essential to developing an effective work plan for the site investigation, and if they are properly formatted, they can provide important quantitative information relating to the hazardous waste.

To use historic map files in conjunction with new aerial surveys, it may be necessary to revise the historic files. The degree of modification required will depend on:

- The CAD format of the drawings
- The coordinate system and projection
- The layering/phase scheme for partitioning files
- Object-naming requirements to link with database tables
- The sizes and quality of the files

If the maps are available only in hard copy, it may be desirable to digitize them to provide a planning tool for locating sampling points or areas of investigation. The effort allocated to this activity should be commensurate with the value received. Alternative approaches are:

- Creating simplified digital drawings for temporary use, containing only limited site features
- Automated digitization of existing drawings
- Use of existing hard copy drawings with acetate overlays of selected new elements such as sampling locations
- Use of existing hard copy drawings until the new aerial survey digitized files have been created

The timing requirements for a specific project and the expected costs should dictate the appropriate approach for proceeding. It is useful for project managers to understand the relevant issues and the most efficient alternatives before proceeding in this area.

4. CHEMICAL LABORATORY ANALYSES

Laboratory analytical data constitutes the largest data set that is compiled during the course of a typical site investigation. Perhaps 60% of the records in a site investigation database consist of the chemical parameters, the measured concentration levels, and their related descriptive information. These data sets usually consist of groups of analyses that comply with EPA standards for conducting site investigations and are defined on the basis of both broad-range screening of samples, and then more focused analyses where specific pollutants are expected or have been identified through earlier screening. The analytical requirements evolve in accordance with regulatory changes that derive from research into toxicity and associated risks, and with improvements in instrumentation and analytical methodologies.

Laboratories contract with site owners or with their environmental contractors to provide a specialized service of delivering analytical results to meet the site investigation requirements. Samples are collected by project field teams and sent to the laboratory, where they are analyzed. The results are submitted in hard copy reports and in electronic format, and contain summary tables describing the concentration levels of compounds as well as a great deal of supporting information describing the analyses and the quality assurance methods that have been applied. Quality assurance sample results are also provided. The formats of these files must be specified clearly if efficient electronic transfer into a project database is to be achieved.

A laboratory information management system (LIMS) may be used to handle the large quantities of data that relate to the sample throughput for which the laboratory is designed. Most large laboratories use a LIMS, either commercially purchased or an in-house management system. LIMSs are designed to accept data from the instruments used in conducting analyses, to reduce the transcription time and errors that can occur through manual handling. The LIMS will have certain data standards for transferring data to electronic format that can be loaded into a summary database. Laboratories may also use manual methods for moving results from instruments into electronic files.

The laboratory results provided by a contractor laboratory are the object of a great deal of anticipation during a site investigation. They are key decision parameters, and project managers need these results quickly and accurately to direct the efforts of field teams who are scheduled to conduct additional sampling, drilling, and well construction on the basis of these results. To facilitate the process, the results should also be provided in electronic format.

When analytical results are obtained for a site investigation, it is important that a quality assurance step be undertaken prior to installing them into a database for broad access and use in making decisions. The QA process at this point includes:

- Reviewing the holding times for the samples, from collection through analysis, to be sure that they are less than the specified values for the specified analysis (aged samples can produce erroneous results and are unacceptable from a regulatory standpoint)
- Comparing results for actual samples with those developed for QA samples, which are designed to evaluate the repeatability of the result, the likelihood of contamination by other samples, the performance of the instruments, etc.
- Reviewing concentration values with respect to reasonableness
- Checking sample identification codes for correctness
- Determining that the records are complete and contain all of the analyses that were specified
- Determining that the record contents (chemical name, qualifier codes, etc.) are correct

The perception that laboratory analytical results consist of a simple list of chemical names and associated concentration levels is far from true. Because these results are so important to the decision process and also may be the subject of legal proceedings, it is essential that all of the important attributes relating to their content be recorded and included in the database in which they are stored. Among the most important of these is a *sample identity*, a code that identifies the sample uniquely so that it can be related to a specific time and location where it was collected. In the course of handling the sample, it may be associated with a number of identity codes by various parties, for example:

- Chain of custody or traffic code assigned by the environmental contractor to define the sample container
- Lot code assigned to a group of samples that are shipped together in the same container
- Batch code assigned to a group of samples that may be analyzed concurrently

These codes relate to the sample and not to the spatial location to which the analytical results will be assigned. The database structure should be clear in specifying a unique *sample location code* that is related to the spatial coordinates where the sample is taken. Sample records should be stored in a sample table that specifies all of the samples (sample identities) that have been taken at that sample location.

Other data elements relating to the analysis are also important to understand and include in the database. Instruments by analytical laboratories have *detection limits* below which their accuracy deteriorates. Concentration values above these limits are meaningful and are generally designated as *hits*. Values can be measured below these limits, but they must be qualified through the attachment of data elements called *value qualifiers*, which describe the quality of a result that may be below the detection limit. The assignment of a result as coming from a *quality control sample* requires the attachment of another value qualifier. The *analytical method* used to process the sample is another data element that is important, since different methods having different levels of interpretation can be used. The *laboratory identity* provides another data element that should be available in the database.

Finally, when interpreting data spatially through the use of GIS and mapping software, the values in the chemical analysis tables provide the inputs to modeling software. Extracting a data set for which a contouring will be applied requires an intelligent selection procedure that may rely on several of these data elements. Contouring should be conducted with consistent data sets (common dates, analytical laboratories, value qualifiers, analytical method, etc.) and can become confounded if data having differing qualifiers or analytical methods are used inconsistently, without a structured and definable selection. When a sample having a nondetect qualifier is encountered, it is important to establish an appropriate treatment of that sample. It can be assigned any value, such as the detection limit or some fraction of the detection limit, at the discretion of the project manager. The value to be assigned as an estimate can be defined on the basis of the qualifier, detection limit, or other parameters in the data table. The value that is used will obviously affect the estimate for the contaminated volume on the site and the subsequent need for further samples, both of which relate the ultimate costs of completing the site investigation and remediation program.

While these points appear to be obvious, experience shows that they can be the subject of a great deal of difficulty in practice. Databases have been developed where it becomes difficult or impossible to precisely relate analytical results to specific spatial and temporal locations. Inconsistency or errors in the database can provide indefensible results. Examples of quality issues that provide a basis for contesting the defensibility of analytical results, which of themselves may be accurate representations of the site characteristics, are:

- Sample location coordinates not related properly to sample identities
- Location codes or coding schemes changing with time as new contractors assume responsibilities
- Missing or erroneous dates associated with sample identities
- Duplication of sample identity from one investigation to the next
- Unknown instrument detection limits

Since the purpose here is to address GIS applications, the focus of concern is with assuring that the sample results can be placed on a drawing of the site, accurately and uniquely, to represent the site conditions properly, and to provide a basis for subsequent statistical representation, modeling, and visualization of the data. Hence, the structure of the *electronic deliverable* that is requested from the laboratory must be consistent with the database needs of the project.

Most laboratories will provide a data diskette that is consistent with *their processing system*; however, the contents may be totally inadequate to the needs of the site investigation team. The specification for the electronic deliverable is critical to achieving efficient data transfer from contractor laboratories who deal with many contractors having diversified needs. Consequently, in developing the specification, the following steps are advisable:

- The specification should be in writing; clearly identifying the format, structure, contents, media, and documentation that is expected with the electronic deliverable
- The deliverable should be contracted, with payment based on correctly provided electronic data in addition to the required hard copy submittals
- A test file should be provided early in the project and evaluated to assure that there will be no delays when production loading is required
- A procedure should be applied to compare the electronic and hard copy submittals as the first step in the laboratory quality assurance process
- Incorrectly prepared files should be called to the attention of the laboratory for resubmission, if the errors are substantive; if corrections are made to the file, the original submittal should be maintained to establish an audit trail for creation of the project database

5. MOBILE FIELD ANALYTICAL SERVICES

To support site investigations, many environmental site investigators use mobile units that provide analytical screening of samples in the field. Typically, a mobile unit would contain transportable instrumentation that can be used to analyze samples quickly in order to provide guidance and direction to subsequent field sampling activities. The samples themselves might be split for analysis by the mobile unit and the remaining portion submitted to a full-service laboratory. In other instances, the mobile unit may be deployed prior to the full-scale sampling program to screen the site and provide input to planning the sampling program.

Field mobile units can conduct sophisticated analyses, but there are limitations to the use of results in a regulatory sense. The data provided by such units are similar to those obtained from an analytical laboratory and can be handled similarly. The submittals from such units do not include the extensive documentation and backup that is inherent in regulated contractor laboratory protocols and are primarily concerned with obtaining concentration levels quickly for each parameter tested. The definition of electronic file transfer requirements and standards can be the same as those for data provided by contractor laboratories.

6. FIELD DATA

The geologic data for a site represent the next important data category associated with site investigations. Such data are collected by field geologists using field logs or data entry forms to store site information and can be available to the project more quickly because laboratory processing is not required. Procedures are needed to assure its availability rapidly and accurately.

As with laboratory data, defensibility of results must be assured. Field geologists use log books to collect observations and data relating to a site when collecting chemical samples, drilling bore holes or wells, or conducting ground water measurements. These books represent the source documents that support the results in a legal context. Other approaches for collecting field data more rapidly include portable computers, voice recognition, and electronic data loggers. If these are used, it is important to understand the legal status of such data by consulting with a lawyer having experience with environmental matters of this type.

The issues associated with moving field data into the project database are similar to those encountered in dealing with laboratory analytical data. However, there are options for how the log book contents are transferred into electronic format. These affect the resources and timing for achieving the objective of making the data available to the project team quickly. Either of the following approaches can be used:

- Coding forms filled out by field team geologists and key-entered by clerical staff
- Key entry by geologists from field logs

Transferring of data onto coding forms is a traditional process for keeping the key-entry activity in the hands of clerical staff who are experienced with rapid key entry; however, this requires additional time and an additional step, which may not be desirable. Key entry of data by field staff can be conducted away from the site, in a field trailer or at the hotel room, after the field work for the day has been completed. This can be achieved with the use of a data entry module within a DBMS system, or through structured spreadsheets that emulate the DBMS. The DBMS may have QA utilities built in that slow down the process but result in better data, while the field teams may be more comfortable with a spreadsheet that has less QA functionality but is easier to use.

An important aspect of field data entry is that field teams must be properly trained in the database structure so that the data are entered properly. There is a significant difference between loading data from a field activity into a geologist's personal spreadsheet, database, or word processor and producing a report; and loading field data into a central database that includes entries from several other field geologists for long-term use by project staff.

A half-day training program in data entry can be an investment having significant value in achieving quality in the contents of the field database. Such a program should include:

- Descriptions of the database structure, table of contents, and fields
- Valid values for sampling locations and other key fields, and other constraints on data entry
- Procedures for submitting copies of data and for maintaining originals
- Data validation and quality assurance reports and procedures for editing data
- Information product types that are available from the data

Standards for data are essential for large site investigations and for organizations having a database containing results from several site investigations. For example, interpretation of the same stratigraphy attributes, color, odor, and texture by two geologists can differ significantly if there are no constraints to the descriptive terminology. The imposition of standards can provide a means for quantifying such information for numerical and graphical display. Comment fields in the database table can store the individual interpretation without constraint to provide the freedom to be creative in describing observations.

Once data have been key-entered, it is desirable to produce output products that assist in interpreting the site conditions. These include:

- Tabular or graphically scaled drilling logs
- Fence diagrams
- Contours
- Cross sections
- Time lines and statistics
- Graphical map products

Creating a *stratigraphic column* or other information product for a specific location can be achieved as soon as the data are loaded and the report or graphic utility required providing the output is available. However, if data representations of geology and chemical results requiring spatial interpretation among several samples or data sets are to be produced, related data sets must be available in the database or mapping system. These would include:

- x,y coordinates and elevations of drilling or sampling locations
- Chemical analysis results
- Base maps
- Two-dimensional topography or digital elevation models

These data sets may not be available at the same time as the field data, precluding immediate development of such products. This is an element of the training program, namely, to provide an understanding of what is needed to produce more sophisticated graphical products. Through coordinated data entry, definition of standardized information products, and training in data management issues, frustrations can be reduced and the rapid production of geological site characteristics can be achieved.

7. AERIAL AND LAND SURVEY DATA

Aerial and land surveys are essential to the completion of comprehensive site investigation and remediation programs. The issues associated with developing quality data from aerial surveys are discussed in Chapter 3. There are additional items of concern that should be addressed when contracting for such services.

Surveyors are registered to perform services in each state and have local knowledge of specific sites within their areas of activity. Aerial firms likewise have experience bases that may or may not include the specific site of concern. Such experience may include access to existing aerial photos or maps and knowledge of benchmarks and foliage patterns and seasonal cycles that affect the scheduling of the survey and the quality of the topographic results. Taking advantage of this experience can provide cost savings and other benefits to the site investigation project.

A telephone interview of the land surveyor and aerial surveyor may be useful to discuss a survey when a site is at a remote location. Aerial surveyors having regional or national presence may have local offices where personal interviews can be conducted. A structured review of the specification should address all of the requirements for providing digital files for the FM/GIS application. The issue of schedule can be very important for a regulated site investigation, and some surveying firms and aerial surveyors may have difficulty meeting the expected delivery milestones.

Land surveyors may serve as prime contractors, subcontracting for aerial survey deliverables with an aerial surveyor. The aerial surveyor may likewise serve as the prime contractor. The prime contractor should integrate all of the deliverables under these conditions. The site investigation contractor or site owner may choose to integrate the two parties as subcontractors, as an alternative operating style. Coordination of survey activities must be conducted carefully by the prime contractor to assure that the resulting products are available in an appropriate digital format and in a timely manner.

B. INFORMATION RESOURCE

It is evident that the site investigation can produce a wealth of information and that there are benefits to managing and integrating this information efficiently. By doing so, it is possible to establish an historic archive of project activities that can be accessible to those who will undertake these steps on future projects. There is a clear need for managers who can control a site investigation and remediation process and complete it economically. The process of integration can result in a graphic archive of the project that includes drawings, statistical figures, photographs, and videos, all accessible through the ES&H system. This is an important organizational resource. It can become:

- The basis for an effective management training program
- An information source for future site actions at the facility
- A public relations tool to communicate a successful cleanup
- A source of information for the organizational strategy regarding hazardous waste sites

III. SITE VISUALIZATION

An important aspect of carrying out the site investigation is providing the investigation team with graphical visualization of the nature and extent of the site conditions. Generic GIS and CAD systems require additional programming or application of drawing functions to manipulate information in order to provide such capabilities. Specialized software tools have been developed to address the requirements of engineers and scientists for subsurface visualization. These tools focus on subsurface modeling and visualization and produce the traditional information products that are applied for these purposes. These include:

* Bore hole stratigraphy drawings
* Cross sections
* Fence diagrams
* Isopleth contours

These have been developed manually or, more recently, with the assistance of CAD. They provide a means for viewing topography, ground water, geology, and contamination plumes in relation to site spatial extent. The mining industry has been a principal contributor to the development of computerized approaches for supporting the development of subsurface views of site data.

A. HIGH- AND LOW-END APPROACHES

With the passage of Superfund legislation, a great deal of research has been conducted to enhance the ability to automate the production of these information products. Further impetus to this technology has been given by military and space technology for numerical surface visualization. As a consequence, a number of software tools have emerged that focus on this specific need.

There are two general approaches to the issue of data visualization:

* Low-end approach using PCs
* High-end approach using engineering workstations

PCs are capable of providing the typical data views identified above, while the engineering workstations have moved toward full three-dimensional volume modeling, including dynamic visualization of subsurface conditions using fly-through, rotation, and animation techniques. These capabilities are dramatic in portraying site conditions to environmental professionals, but they are also understandable to nontechnical managers, lawyers, and judges. The value of this effect in litigation is obvious.

Industrial organizations and regulated government agencies have often adopted the PC approach, primarily due to cost differences. Federal and state regulatory agencies have begun to adopt the high-end technology, which has been effective in presenting their cases in litigation against industrial organizations. Both approaches have their place in the context of site investigations. The differences are:

- The degree of visualization possible
- The flexibility in changing views
- The quality of the on-screen graphics
- The quality of hard copy information products

High-end products are available that provide:

- Full views of all numerical surfaces with transparency or solid representations
- Photo overlays of surface topography
- Sophisticated gridding algorithms designed to handle plume volume generation in three-dimensional space
- Three-dimensional building representation in conjunction with subsurface views
- Dynamic rotation or movement of observer and view point
- Three-dimensional on-screen sectioning through topography, ground water, geology, and contamination plume

To achieve these capabilities, powerful engineering workstations are used having large quantities of memory (32 MB) and disk storage (1 GB), and high-speed numerical and graphical processors designed specifically for three-dimensional graphical modeling. Silicon Graphics workstations are typical to these applications, while Digital Alpha is emerging as a viable alternative. Hewlett-Packard, Sun, and VAX workstations are being used in intermediate applications, having some of the above capabilities to a lesser degree, but clearly superior to the low-end PC applications in performing these functions.

The data structure for the advanced applications is generally quite simple, consisting of four-dimensional arrays (x,y,z coordinates and parameter to be displayed). The need for high-powered capability is evident from the memory requirements. A typical example would be:

- Original data set (x,y,z,p) of sample values = 15 KB
- Gridded data set representing volume = 150 KB
- Graphical data set for on-screen volume viewing = 15 MB

This growth by a factor of 1000 clearly indicates the resource price to be paid for achieving the visualization functionality. Due to the limitations in processing such large data sets, PCs are usually not a viable platform for such activities. However, the dramatic growth in PC technology may provide capabilities of this type in the future. On the other hand, the workstation vendors continue to improve in price/performance along with PCs. As a consequence, the three-dimensional visualization technology should be viewed as a specialty application to run on a limited number of workstations rather than a broad application over hundreds or thousands of PCs in a full enterprise implementation.

It is useful to note that three-dimensional visualization applies to gridding and modeling techniques and is faced with the same uncertainties and possible errors that are experienced in two-dimensional gridding. The ability to visualize provides a powerful management communications tool; however, unless the results are based on sound and defensible mathematical methods, they can be attacked as unreliable.

B. SOFTWARE TOOLS FOR VISUALIZATION

There are a number of software tools currently being applied to site visualization. Examples of a number of these are:

- Site Planner — an intermediate product by ConSolve that runs on Sun workstations (and PCs with limited functionality) and has increased data-viewing capabilities
- GIS/Key — a FoxBase database structure with AutoCADtype functionality for displaying data on PCs
- Stratifact — a FoxPro database with stratigraphy, cross section, and contouring capabilities running on PCs
- Earthvision — a product by Dynamic Graphics that runs on Silicon Graphics and VAX Alpha workstations and provides extensive data visualization
- FMC Ground Systems — a service by FMC running on Silicon Graphics workstations for subsurface visualization
- Advanced Visualization System — a generic visualization system emerging from medical and manufacturing applications that is also being applied to environmental graphics
- *EDGE* — Environmental Data Graphics from ERM, an integrated FM/GIS prototype and development tool for environmental facility management, which employs techniques for visualization of stored data

While this is not an exhaustive set of all of the tools being used for visualization purposes, it is representative of the technologies available at the time of this writing. The explosion of information technology is such that these tools will evolve to incorporate the hardware and software capabilities available on their selected platforms. Examples of the graphical outputs produced by these types of systems are provided in Section IV in Chapter 5.

IV. SITE REMEDIATION

The ultimate objective of a site investigation is to ascertain the appropriate remedial action to be taken on the site. This may consist of no action and continued monitoring; removal of contaminated material; or large-scale remediation such as pumping and treatment of ground water, or on-site treatment using chemical, incineration, or biological processes after excavation or *in situ.*

A. ENGINEERING FEASIBILITY STUDIES

The process begins with an engineering feasibility study (FS), which evaluates alternatives for dealing with the site and with the associated costs of these alternatives. FS activities are usually conducted in conjunction with, or shortly after, the site investigation. They must rely on the results of the site investigations to provide engineering variables that are used in the study. The design issues can be quite complex and relate to a variety of variable types, e.g.:

- Extent, volume, shape, and stratigraphy of the zone to be treated based on alternative cleanup levels
- Geotechnical conditions describing the structural characteristics of the rock, soils, and soil stability angles
- Flow characteristics of ground water under natural conditions and during pumping
- Combustibility of soils and rock and characteristics of their ash products
- Behavior of the water or soil with regard to the treatment alternatives (treatability)
- Costs associated with treatment alternatives

The volume and shape of the excavation can be studied by using the CAD component of the ES&H system in conjunction with the site investigation database. Geotechnical and treatability parameters can be stored in look-up tables and related to the stratigraphy data and the subsurface characteristics compiled during the site investigation and stored in the ES&H data tables.

Information and models are available in the public domain and can be used to support engineering studies and remediation planning. For example:

- The EPA Vendor Information System for Innovative Treatment Technologies (VISITT) provides information on new treatment technologies, describing unit costs, site conditions where they are being applied, and actual site locations. This information is available on diskette and can be used to estimate expected costs, based on experience from comparable sites. It could be linked to and become a part of the ES&H system functionality.
- Remediation cost models developed by federal government agencies such as the EPA and the U. S. Army Corps of Engineers can also become part of the remediation management component of the system. These models

could be linked to the ES&H system data tables by delivering and formatting site-specific data in accordance with the model's required input files.

- USGS offers digitized map files that can be used to obtain a preliminary level of topographic information, prior to more detailed site mapping.
- Ground water models are also available in the public domain, and their input data relies on site conditions, as defined by chemical concentrations and soil and water characteristics stored in ES&H system tables.
- Air dispersion models such as ISC2 can produce concentrations and risk isopleths as outputs, and these results can be displayed as site drawing layers, or as time series curves.
- The EPA Air Chief (Clearinghouse for Inventories and Emission Factors) is a CD-ROM system developed to provide emissions data available from EPA reports and databases.

These and other available computational aids and information sources can be integrated into the ES&H system to support the FS process. Database information and drawings transferred electronically to workstations used by engineers in conducting feasibility studies can be an important step in the structured evolution of the ES&H system. It is important for organizations to develop in-house focus centers that maintain awareness of the emerging information sources and make them available.

By integrating these steps, it is possible to preclude the problems that were identified earlier regarding data inconsistency. The information acquired during earlier phases of the project should be readily available for use in subsequent phases. The information used during the FS and design phases should be integrated with the RI phase information in an efficient project life cycle. The requirement for electronic deliverables that applies to an analytical laboratory during the site investigation should not be unique to that step. Site investigation results should be delivered electronically to the engineers conducting the feasibility studies, and likewise for the remedial design. These steps can be achieved through:

- Integrating the appropriate models and FS support tools within the ES&H system environment
- Networking the engineering workstations or PCs where the FS analysis and modeling is being conducted with the ES&H database and drawings
- Establishing standards for file transfer between the FS computers and the ES&H system that can be used with or without networking

These approaches define levels of integration that facilitate the process and reduce the time and costs required to complete the FS activities. Their implementation demands require structure and standards to interface information across the process steps that allow the information to flow with less resistance.

B. REMEDIAL DESIGN

The design of site cleanup requires another set of steps that are based on the results of the engineering feasibility studies. Designs are documented by engineering computations and drawings. The drawings must be produced at construction accuracy levels to support construction activities. If site drawings at these levels of accuracy are not available during the site investigation, additional aerial or land surveys may be required.

When a site owner conducts the investigation and design phases or acquires the services of a single contractor to conduct these activities, there is every opportunity to be efficient in the flow of information from one phase to the next. If two or more contractors are involved, the site owner should establish standards for information exchange between the contractor's data and drawings and the site owner's ES&H system. It is also important to assure that information flow among contractors is efficient (since the site owner is paying for these services). These drawings are, in essence, a part of the facility management process and will represent the facility in the new, postremedial state. Importing them into the system should be as systematic as all other aspects of data and drawing management that relate to the system.

Finally, automated excavation design and volumetric analysis are useful components of the ES&H system functionality. They provide an organization with a means for evaluating designs independently of their contractors, using the same data and drawings that are available in the ES&H system. The costs of site cleanup are significant and depend on the results of the design process. The ES&H system can provide an effective quality assurance step to validate designs and control costs. A rapid and efficient independent analysis can be conducted by an organization's engineering staff to assure that designs provided by contractors are reasonable. The purpose of the system is to *manage the facility* from an ES&H standpoint. This means that managers should have access to information and, in certain areas, know as much as or more about their facilities than their contractors. Better management of contractor activities can be achieved by taking advantage of the available information and integrating it effectively.

C. CONSTRUCTION

During the construction phase of a remediation program there is a continuing need to maintain management oversight and control of the process. The construction status can be illustrated graphically through Gantt charts, expenditure curves, and earned value summaries based on project schedules and budgets stored in the ES&H system. These can also be portrayed by sketches or color-coded overlays on drawings illustrating the area expected to be completed on a given date in comparison to the actual area. The status can be communicated quickly and effectively to managers through these graphical representations to monitor current conditions at the site. Graphical displays of site conditions during construction can be transmitted from the site via electronic mail and viewed by managers in *view only* mode. Photographs and videos of site activities can likewise be viewed electronically to facilitate information exchange.

Chapter 5

APPROACH FOR AN ES&H FACILITY MANAGEMENT SYSTEM

This chapter describes an ES&H information system prototype that integrates databases, maps, drawings, and regulations to provide the ES&H Manager with a powerful tool for data access and visualization. The relationships between the database objects to database tables and key fields are described in the context of specific applications.

The preceding discussions dealt with the data types and issues faced in developing information that can be used to support ES&H applications. MIS departments are charged with developing the organization's information resources to meet the need for supporting ES&H requirements. It is also important for ES&H managers to understand the resources and technologies that are available and to be sure that the appropriate ones are being used. The principal concern is in the preparation of information products that can support management's needs.

I. INFORMATION PRODUCTS

Information products of various types are needed by managers who deal with ES&H issues. There is often a great deal of frustration with the large quantities of data and the sparse amount of meaningful information. There are many reasons for this frustration and a few of these are:

- ES&H managers are not able to express their needs precisely
- MIS departments often focus on information technologies; and their staff may have little understanding of the ES&H requirements
- Point solutions (implemented to address specific needs) are not integrated; and the requests for information may require data located in two or more disjoint systems
- Drawn-out development efforts require extensive coordination among parties; and often these efforts do not come to satisfactory closure
- Resources and costs required to implement an effective system are underestimated and lowest-price approaches are often implemented

- The frequency of regulatory changes imposes a continuing need for under-standing new requirements, which exacerbates all of the above conditions
- The information technology explosion creates an atmosphere of indecision as new products and systems promise a better approach tomorrow, with the result that investments in today's technologies are delayed

A. INFORMATION PRODUCT TYPES

While there are no easy solutions to these conditions, adopting a rational strategy for effective information management is essential. The strategy should address the need for different types of information product types, such as:

- Compliance reports for external submission that must be provided to regulatory agencies in accordance with annually defined cycles
- Management reports for internal use to support periodic requirements for compliance
- Planning reports that are used within the organization to measure perfor-mance against ES&H metrics, to manage the expenditure of resources, and to disseminate performance with annual financial reports or for public information purposes
- Unforeseen requests for information in response to incidents or emergencies

B. VISUALIZATION

An important aspect of ES&H information requirements is that they inher-ently have a spatial context. Spills, emissions, permits, samples, contamination plumes, accidents, chemical inventories, work places, and cleanups all have associated locations. These can be represented by a tabular address or code, but intuitive understanding is enhanced when they are displayed graphically on a map, drawing, or figure. The growth in windows-based computing, which permits multiprocessing in a visual environment and makes it easy to monitor one's selections and steps, is the result of the receptiveness of the user commu-nity to this intuitive and natural manner for interacting with information.

GIS technology provides the intuitive link between databases and the graphical framework to which they can be attached. The future of ES&H systems resides with GIS technology, and managers in these organizations as well as MIS departments that support their needs must develop strategies for moving in this direction.

Information products can be developed using GIS and integrated with products that are available through multimedia technology. The typical types of information products that are required in these applications are:

- Tabular lists of information available from traditional computerized data management technology
- Hard-copy representations of site information displayed on a map or plant drawing
- Screen images displaying photographs of plant site conditions that can be displayed in group meetings to provide for focused problem solving or decisions
- Graphical files and numerical surfaces that can be used in conjunction with enhanced visualization software for spatial displays of waste site problems
- Dynamic queries of databases to ask ad hoc questions and translate the answers into changes on maps or drawings to illustrate *where* and *how much*
- Pointing at maps, plant drawings, schematic drawings, sampling locations, emission points, or waste sites to provide data, tables, graphs, photos, or statistics
- Dynamic visualization of plant conditions through the use of technologies that allow rotation and observation of surface and subsurface characteristics and *fly-through* displays that simulate movement over and through terrain

II. PROTOTYPE SYSTEM

The issues that environmental organizations face with regard to compliance and site remediation are described in earlier chapters. The purpose here is to describe the functionality of an integrated system that has been developed for this purpose. The system itself is a pilot project that has been developed for two purposes:

- To demonstrate the integration and accessibility of data in an environmental FM context
- To manage data for large-scale site investigations

The system developed by ERM, called *EDGE*, has evolved from ES&H applications. It will continue to evolve as new requirements and technologies become available. It has been implemented within a specific hardware/software environment using commercial software systems and tools so that it can be installed readily within this environment. Moreover, the prototype has also provided a learning platform through which development within or migration to other hardware/software environments can be accomplished. To that end, *EDGE* is an approach to integration of technologies within a GIS framework for use in managing information for ES&H applications. It is an example of the principles of structured evolution.

A. SYSTEM STRUCTURE

The *EDGE* approach applies three principles in addressing organizational ES&H needs:

- A strong graphical user interface (GUI) — an object-based GIS serves as the principal user interface within a windows environment to provide conventional GUI as well as *geo-GUI*. (geographic GUI)
- Total environmental integration — environmental compliance and remediation are addressed within the same data and drawing structure, and safety and health systems can follow within this same framework
- Preservation of existing organizational systems — links to commercial environmental databases, networks, documents, and drawings are established for this purpose

1. HARDWARE AND SOFTWARE

The hardware and software components integrated within *EDGE* are:

- Engineering workstation for processing functions
- Network to PC to view drawings
- Object-based CAD/GIS for drawing functions
- Site modeler for site investigation and remediation
- Structured query language (SQL) RDBMS for tables
- Commercial compliance DBMS ENFLEX® (non-SQL) for stand-alone use and interface with GIS
- PC emulation on the workstation
- Image viewer for documents and photos
- Remote access via windows terminals or PC terminal emulation over a network to view drawings

Figure 5-1 illustrates the structure of the *EDGE* system, which integrates regulatory and remediation database tables with facility graphics to produce information products. The system has been used to demonstrate the capability of integrating a variety of software tools on a single workstation platform and for providing access to users at remote locations. It emphasizes ease of interaction with the data through a geographic GUI. These concepts have been implemented on workstation technology, before personal computers had the capability for integration at this level. They can now be implemented, to a great extent, using PCs.

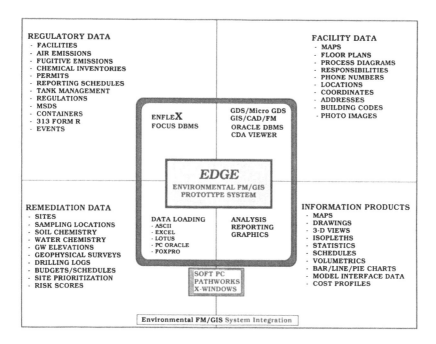

FIGURE 5-1. *EDGE* prototype environmental FM/GIS system.

2. GRAPHICAL USER INTERFACE

The geo-GUI consists of a spatial and object-based interface with on-screen graphical representations of the organization's facility. This starts with a map of the region where the facilities are located, which can be the U.S., a group of states, a foreign country, or the world. Facilities are identified by geometric objects or symbols placed on the screen at their approximate geographic locations. At this level, geographic precision is not required. The locations are overlaid on a base map representing the geography of the appropriate area. The names of locations can be displayed permanently or on demand, along with other data stored as attributes within the GIS, and recallable through menu selections and pointing at the selected facility location.

The "windows" environment provides concurrent access to a number of integrated processes. Typically, there may be base maps or process drawings, database tables, graphs, images, menus, and reports available in a windows setting at any point in the application. The typical screen layout includes a graphics area, a menu area, a dialog area, and a drawing display area. The screen layout can be modified by the user by invoking windows functions to provide a structure that may be more appropriate to a particular application or functionality.

The base map for each facility is stored in CAD format within the system. Base maps are imported from different CAD systems and stored in a structure layer format. Through menu and facility selection commands, a base map for a selected facility may be recalled. Subsequent commands provide a means for modifying the drawing to include or exclude specific layers that are appropriate to subsequent investigations or reporting requirements.

Interface with the base map and main menu produces additional menus and functionality for accessing, analyzing, and viewing environmental data in relation to the base map features. The pop-up menus appear, based on selections from the main menu, dialog prompts, or clicking on objects in maps or drawings displayed in the graphical area.

B. SYSTEM FUNCTIONS

Since the system is designed to address both compliance and site remediation issues, menus are designed to proceed into areas that support these activities. The menus are organized to activate a number of general functions programmed in fourth generation languages and can be used to provide greater flexibility and information product variety. The following general functions are available:

- Access facility-level information
- Manipulate drawings
- Access and display database tables
- Access to and interface with a commercial compliance DBMS
- Display and model database contents on drawings
- Spatial modeling and visualization
- Access and display images related to locations
- Construct menu-driven queries that result in graphical responses
- Develop cost- and schedule-based reports and graphs for environmental compliance and remediation
- Use numerical surfaces to support site investigation and remedial construction
- Executive information system access to the data

1. FACILITY-LEVEL INFORMATION

For each facility located on the area map, facility data are stored as attributes within the FM/GIS, and selected attributes can be displayed in a dynamically generated report through menu clicks. In addition, a facility database table or a facility report can be displayed from a database table storing data for any facility that is located on the area map through the point-and-click procedure. The facility report uses a standard reporting format imbedded in a commercial environmental software package that is linked to the facility objects on the base map.

2. DRAWING MANIPULATION

Once a facility base map is selected, it can be manipulated through the application of three functional approaches:

- The use of menu commands to change its format
- The use of icons to apply CAD zoom commands
- The use of button commands for CAD queries, redrawing, or other such commands
- Through direct key-entered commands in the dialog area using the native GIS macro or programming language

This approach provides easy access to the system for the new and interme-diate-level user and the flexibility to use the full FM/GIS capabilities without exiting the geo-GUI interface. This latter feature is important to the preparation of complex or composite drawings that might include several components, e.g., several cross sections along with a base map (including contours), all displayed concurrently as a screen image or as a hard-copy graphic.

3. DRAWING ACCESS

Drawings of several types are included in the system, including facility maps, building drawings, and process flow diagrams. A site map is accessed by selecting the facility on the area map. Selecting a building on the facility drawing produces options for viewing any floor of the building in greater detail, and includes process drawing objects at their appropriate locations in the buildings. Detailed process drawings are displayed by pointing at these objects. These drawings are automatically drawn through the simple point-and-click process. Once they are displayed, they can be modified or used for geo-GUI purposes or for links to database tables and analysis functions.

4. DATABASE ACCESS

Each drawing is enriched with object intelligence, making it possible to access the database tables through a number of methods, quite similar to the process for manipulating drawings. These include:

- Program execution accessing the DBMS to produce a data screen based on selection of the object. For example, by pointing at a sampling location, the chemical results table is displayed, and all sampling results at any depth can be produced by scrolling through the table. The same approach applies to air emissions samples and associated emissions results.
- Data extraction from database tables to produce summary reports or graphics, or to post data on a graphic. For example, selecting a sampling location from a drawing and the depth and chemical name from a menu, the database sampling results are posted at the appropriate sampling location.
- Direct access to the database through the use of the native DBMS language. An SQL command can be key-entered in the dialog area to extract data from any table and format it for reporting or subsequent analysis.

As with drawings, this provides flexibility for differing levels of user proficiency. A variety of menu-driven options are available for standardized tables and data views, but the capability of extracting data from any table and using it in conjunction with any drawing is also available.

5. COMMERCIAL COMPLIANCE SOFTWARE INTEGRATION

Another aspect of system integration is demonstrated through the use of commercial software for environmental information management. ENFLEX DATA® is an example of a commercial environmental software system using hierarchical DBMS technology and the FOCUS DBMS. Some organizations that may have ENFLEX DATA® or other commercial software cooperating on several PCs in a distributed environment may want to integrate this system within an ES&H FM. Using the principle of structured evolution, the approach for system integration within the *EDGE* framework is described in the following steps:

- Porting the ENFLEX DATA® modules to a minicomputer environment for multiuser access
- Installing the modules on the *EDGE* workstation (having the same architecture as the minicomputer)
- Providing independent execution of ENFLEX® modules and the hierarchical DBMS within a separate window under *EDGE*
- Linking selected modules through data table location codes for facilities and areas where chemicals are stored to drawing objects representing these same locations
- Developing menus and location interfaces within *EDGE* to auto-execute the modules based on geo-GUI selections
- Providing access to the module tables and reports within *EDGE*

This pilot project demonstrated a means for integrating a non-SQL DBMS within *EDGE*. This provides credibility to the approach and takes an important step in the evolution process. The pilot system demonstrated the feasibility of interfacing the ES&H applications and geo-GUI with any existing DBMS that is portable to the environment.

The pilot system was used to demonstrate the feasibility to users of the commercial compliance system. This established requirements for the next phase of evolution:

- Reviewing available plant CAD drawings for future integration
- Developing a scope and costs for a pilot project to implement existing facility drawings and data within the *EDGE* framework

At this stage, the alternatives available in the structured evolution path can be defined, e.g.:

- Converting the hierarchical data structure of the modules to the *EDGE*'s RDBMS to strengthen the integration and rewriting the screens and reports, or
- Maintaining the hierarchical system and implementing full linkages of all modules, or
- Operating in a dual DBMS environment for some time period with evolution to a client-server environment.

In proceeding with the evolution, it would be necessary to address issues such as:

- Selecting the modules, data sets, and facilities to be used
- Importing drawings and adding intelligence to them
- Creating an operational pilot implementation for a selected set of users
- Evaluating the operational pilot based on user experience
- Presenting the results of user experience to the organization's management, and demonstrating the operational pilot system for them
- Establishing the appropriate next steps

6. SPATIAL DATA MODELING AND VISUALIZATION

Spatial modeling is an important aspect of the GIS capabilities in providing visual representations of site conditions. Such modeling can be conducted through the use of data that are stored in drawing format, such as topography; database table contents, such as chemical concentrations at sampling locations; and combinations of the above to produce more complex spatial representations of site conditions.

A numerical surface is a set of x,y,z values that can be used to represent topography or ground water elevations, for example. Such surfaces can be viewed as contours in a two-dimensional plan view or as three-dimensional surfaces. Through the use of more sophisticated solid modeling, they can be rendered to provide enhanced representations that are more visually striking.

The GDS Site Modeler utility is used to produce a variety of graphical products that are useful to the site investigation process. These include:

- Menu-driven contouring using the triangulation method
- Data thresholding to identify areas of contamination associated with se- lected cleanup levels and direct computation of associated areas
- Three-dimensional surface representations in contour or triangulated ir- regular network (TIN) model representation
- Generation of cross sections through topography, ground water, or other surfaces
- Volumetric computations of differences between surfaces, which is useful in planning remediation

To extend visualization capabilities, it is sometimes useful to apply addi- tional capabilities to provide insight into site conditions. The GDS Solid Modeler utility within the GIS provides a means for enhancing the visual impact of the available surfaces. These capabilities include rendered represen- tations and intersections of multiple surfaces, e.g., contamination plume and ground water, and the use of colors and shading to accentuate the visual impact of the rendering. A preferred alternative that is being applied for visualization is to export the topography, ground water, stratigraphy, and chemical concen- tration data from *EDGE* as sets of x,y,z,p (parameter) files to specialized visualization software tools that are designed specifically for this purpose.

7. IMAGE ACCESS

Access to photographs and documents related to the facility and to specific objects on the facility drawing is another important requirement of ES&H information management. Images must be organized for access. Image infor- mation is stored in a database that provides GDS with the name, description, location, and format of the image. The system provides a menu selection that executes an imaging software system, which in turn selects the appropriate image that is linked to specific objects on a drawing. For example, a material safety data sheet can be displayed for a particular building location or process; a photograph of a waste site or of site investigation activities can be displayed by clicking on the building or waste site. This initiates the execution of the imaging software within a new window, extracts the selected image, and displays it in that window. Access to the full functionality of the imaging software for loading images, indexing them to locations, etc., is also available at that point. Integration of imaging technology within the geo-GUI is another intuitively based approach to information access that is finding use in the ES&H arena.

8. COST AND SCHEDULE DATA AND GRAPHICS

The schedules and budgets for environmental activities constitute another information component that are usually disconnected from the technical data. The *EDGE* approach integrates this data through the development of tables within the DBMS that are structured to facilitate their linkage to GIS-based location entities. The structure includes:

- A work breakdown structure (WBS), for time-phased activities
- Cost account codes
- Hierarchical roll-up from WBS codes to sites, facilities, and the total organization

In this manner, the user can select reports or graphical representations from the financial tables in the database at any of these levels. The data can be aggregated at the corporate, site, WBS, or cost account level to produce bar graphs, pie charts, or time lines of planned and actual expenditures and schedule status. As with technical data, the geo-GUI provides access to these reports and graphics by pointing at pop-up menus and related drawing objects (facilities, sites) to which the WBS applies.

9. CHANGING DRAWING FEATURES THROUGH LOGICAL QUERIES

The interface between the intelligent drawing and the database is best exemplified through the use of logical queries against database tables that produce responses graphically through changes in drawing features. The queries are constructed through a series of menu selections to construct commands, define or enter variables, and select logical operators; for example:

- Select facilities from the area drawing and highlight those where budgets exceed a specified amount, or where expenditure rates exceed budgets
- Select fugitive emission points on a process drawing and highlight the equipment components where emission rates exceed specified levels, or those leak rates have not been reduced in accordance with a specified standard
- Select and identify sampling locations where certain contamination criteria are met

After the query is made, the map or drawing features change through
highlighting or the appearance of symbols at those locations that pass the
selection criteria. The tabular list of identification data that would normally
result from such a query in a traditional DBMS context can simultaneously
be displayed in the dialog area of the GIS. These steps set the stage for
invoking subsequent commands, such as clicking on the selected objects
(facilities, equipment, sampling locations) to activate and review specific
data elements, or to produce graphics or model the data that relates to the
selected entities.

10. SITE REMEDIATION DESIGN

The remedial design of a waste site is a natural extension of the site investigation process and relies on the same data sets. All too often, the remediation
process begins with the need for a whole new set of data and drawings. This is
because the site investigation system may not be at the proper level of precision
for design or it may not have the appropriate design functionality.

The *EDGE* approach includes full CAD design functionality for the
remediation process, which is implemented through menu-driven preliminary
designs. These provide the capability to complete the preliminary design in a
few minutes through:

- Using the contamination data to provide a site profile around which excavation can be designed
- Constructing cut surfaces and patching them to produce an excavation
  design
- Computing excavation volume based on the design parameters and relating
  them to expected costs
- Viewing the excavation from arbitrary site locations and elevations to
  visualize the change in topography

These steps are menu-driven and utilize the same CAD and database
components that have been employed during the site investigation. The responsiveness of the system provides extensive flexibility in revising design parameters and producing alternative designs and associated costs quickly for use in
planning remediation and in establishing negotiating positions with regulatory
agencies.

C. INTEGRATING WITH SPECIALIZED SOFTWARE

A GIS is a generic tool that is intended to provide linkages between maps
and data and to process and analyze graphical objects for a wide variety of

purposes. The *EDGE* approach illustrates how the tool can be customized to ES&H applications for both FM and site investigations. There are a number of specialized software products in the market that address one or the other of these areas, and some of these have been selected by organizations for use at selected facilities or for broader organizational purposes. The *EDGE* approach fosters integration with such systems when a corporate advantage accrues. For example, data sets have been exported to visualization software to provide high-resolution subsurface graphical displays of topography, ground water, geology, and contamination, using the power of specialized tools for this purpose. No single software tool can encompass all of the requirements of an effective environmental facility management system. Integration with available technologies and with specialized tools is the appropriate strategy for system evolution.

An example of such integration has been demonstrated by Maitin and Klaber in a paper entitled "Geographic Information Systems as a Tool for Integrating Air Dispersion Modeling". This paper explores the integration of the industrial source complex (ISC2) model within a GIS for contouring and display of concentration and risk contours. This approach has been used to integrate PC applications and to link them with a VAX workstation using a Pathworks network. It provides an excellent example of the use of public access modeling software in conjunction with GIS and database systems.

1. REQUIREMENTS AND ALTERNATIVES

Staff in ES&H organizations and in the consulting organizations that service them have adopted tools to support their activities. These include spreadsheets, databases, statistical software, contouring and numerical surface software, and more sophisticated systems that relate to specific applications. Among these are geologically focused software systems for developing fence diagrams, bore hole logs, and other graphics to support analysis and understanding of subsurface conditions. Integrating with these tools may be a critical requirement of a successful implementation of an FM/GIS system. They can represent a corporate investment that should be preserved, not necessarily because of the direct software costs, but because of the existing knowledge base and procedures that are based on these systems.

Integration with such systems can usually be accomplished in varying degrees, depending on the ease of use that is desired and the user transparency that is required. Two primary types of integration are:

- Loose coupling — file transfer
  - Diskette transfers of data and drawing files from the ES&H database tables to the specific application using defined ASCII data structures
  - LAN or WAN interface to transfer the ASCII files more efficiently

- Process control — GIS control of the RDBMS
  - Use of a common database in a client-server environment to remove the need for special ASCII file creation
  - Full integration of both applications under a common interface, single database, or GUI

The specialized systems that are the subject of this discussion have differing structures and, therefore, differing capacities for integration. The *EDGE* approach has been to move toward the highest level of integration wherever possible, and to adopt lower levels where necessary. If a specialized system is in use or is being contemplated for acquisition, the organization's information management group should clearly define the level of integration that will be available to users who may be attracted to the functionality and may not realize the integration requirements. While the movement to open systems is the direction of information technology, there will be a need to address the specific integration requirements of linking today's installed base with other preferred systems for some time.

2. EVALUATION OF ALTERNATIVES

Establishing the appropriate integration approach is best achieved by defining the requirements and then proceeding with a structured evaluation procedure, which includes evaluation steps, criteria of performance, and documentation of the results. For example:

- Evaluation steps
  - Install software
  - Install test drawings and data sets
  - Identify functions to be tested
  - Schedule test steps
  - Conduct tests
  - Prepare conclusions
  - Review conclusions with users
  - Make recommendations
- Evaluation criteria
  - Ease of transfer of drawings
  - Ease of transfer of data tables
  - Ease of preparation of specialized information products
  - Extent of capabilities
  - Quality of on-screen graphics
  - Quality of hard-copy graphics
  - Openness of software to integration
  - Consistency with organization's software/hardware directions

- • Costs for software and hardware
- • Maintenance and support availability
- • Training requirements and costs
- Costs for integration and implementation

Documentation of the process and the criteria can be accomplished using the above elements. If several competing systems are being evaluated, the criteria can be prioritized, assigning a higher weight to more important criteria, and scored, perhaps, with a value from one to five. Composite scores can be prepared to provide relative comparisons more readily. Decisions can then be made with a documented basis of support.

III. EXECUTIVE INFORMATION SYSTEMS

The data variety and extent within the *EDGE* approach provide an important level of visibility at the organization's executive level. The same databases that are available to engineering and scientific staff in dealing with ES&H compliance and remediation issues also can have utility in an executive information system (EIS). These systems are used for strategic purposes such as:

- • Establishing metrics that convey quantitative measurement of ES&H performance against goals
- • Prioritizing resource expenditures
- • Planning and management of expenditures
- • Obtaining visibility into diverse and geographically distributed operations

The level of detail of information that is made available to executive managers for these purposes can be left to the discretion of the organization. They can rely on existing information systems or they can implement independently derived data or summary information.

A. APPROACH

Executive information systems should provide rapid visibility into current conditions if a facility- or site-specific issue is raised to a high level of awareness. They should also provide summary information concerning expected costs and potential liabilities that may result at corporate sites. The *EDGE* approach is effective in meeting the need for making appropriate information available at the executive level through the same intuitive geo-GUI that is applicable to engineering and scientific uses.

Extending the information access to these higher management levels can be accomplished by:

- Implementing specialized applications for executive use
- Summarizing *EDGE* information for executive view
- Providing access to selected components of detailed information
- Providing full access to *EDGE* functions
- Combinations of these alternatives

An application that has been implemented for executive use relates to the management of site remediation budgets for an organization having a large number (about 150) of waste site activities. The facilities are located on a map of the U.S., providing the same entry point as for engineering and scientific uses. The information access functions that originate from this point are:

- Converting the locations to color codes in accordance with business unit or other organizational identities
- Implementing specialized database tables to store financial, legal, regulatory, and public image variables
- Implementing prioritization and decision criteria that use these special tables along with data elements from the basis of engineering and scientific data tables
- Screening the facilities using the prioritization criteria to identify those that are of specific concern in a spatial, geographic context
- Selecting a high-priority facility and reviewing data and special summary reports
- Displaying the base map for the facility and reviewing drawings, images, and data that relate to the prioritization issues

This top-down approach is effective in bringing the executive into contact with facility information in a manner that can be very appealing. The executive is engaged in understanding the problems and can do so without the need for volumes of documentation and maps that would otherwise be required. The data and drawings that are being viewed are the same as those being used by the facility, engineering, and ES&H groups to conduct their activities. Communications are enhanced and delays in access to important information are reduced. With appropriate communications networks, plant personnel at remote locations and executives in corporate headquarters can review the same graphical information simultaneously to improve the effectiveness of teleconferencing.

IV. EXAMPLE INFORMATION PRODUCTS

The figures on the following pages are representations of information products that have been produced to address the above needs. They include graphical outputs from high-end workstation visualization systems, from PC-based systems, and from an integrated environmental facility management prototype. Such systems will become commonly used within organizations to provide visibility into the ES&H issues affecting their facilities. Making them work requires a *structured evolution* from current computing environments and systems, one that preserves the current investment and moves toward new technology through orderly steps, setting appropriate expectations, and achieving measurable progress.

A. VISUALIZATION OF CONTAMINATION PLUMES USING HIGH-END WORKSTATIONS

The subsurface viewing of contamination plumes is, perhaps, the most dramatic application of visualization technology in an environmental setting. Software vendors have been asked to provide examples of their products, and a sampling of these are provided on the following pages.

Plate 5-1,* from Dynamic Graphics Earthvision, was prepared using a Silicon Graphics workstation. It represents a chair-cut cross section (cut along x and y axes) through a site in Virginia, illustrating contamination in ground water ranging from 50 to 10,000 ppb beneath a surface represented by a USGS quad map. Plate 5-2, also from Earthvision, illustrates another chair cut through the Vadose zone and aquifer with the surface features in transparency. Each of these graphics can be revised on screen through menu selections, and represent high-end visualization functionality. Data are imported in four-dimensional (point/parameter: x,y,z,p) tables and are modeled using the minimum surface tension algorithm. The resulting subsurface graphics can be displayed on-screen in static views or in time-sequenced dynamic graphics employing a proprietary algorithm. Viewing location and view direction can be changed to provide "fly-through" type emulation.

A second high-end visualization system is the FMC visualization system, which is used to support computer-based analysis of spatial data. This is an example of military technology transfer to civilian applications. The system also runs on Silicon Graphics workstations. Plate 5-3 from FMC illustrates a section through a contamination plume on and beneath a gridded topographic surface. Well locations and surface features are also displayed. Plate 5-4, also from FMC, displays surface features in conjunction with well locations and subsurface contamination.

Another high-end visualization and environmental modeling capability called MGE (modular GIS environment) comes from Intergraph. The MGE Voxcel Analyst employs a modeling method that uses hexahedral (six-sided) voxels (three-dimensional extension of the pixel concept) to represent subsurface features. The voxels can be uniform (cubic) or nonregular in shape and spacing to conform to physical or geological formations. Plate 5-5 illustrates

* Plates 5-1 to 5-9 appear after page 82.

an example of model outputs from MGE, which is similar in appearance to the previous two high-end visualization systems. The difference between MGE and the above systems is that it incorporates a full set of GIS functions as well as the visualization capability, and integrates with a relational DBMS in addition to importing ASCII files for modeling. The voxel capability runs on proprietary Intergraph workstations, while a number of MGE modules run within the PC Microstation version of the Intergraph system. A port of MGE to Windows NT has been initiated with releases of some modules during 1994.

B. OTHER VISUALIZATION SYSTEMS

In addition to the high-end visualization systems, there are intermediate systems running on Sun, VAX, and HP workstations, and others running on PCs. The ConSolve Site Planner uses the term *virtual site* (not a virtual reality capability) to address site characterization and remediation. Site Planner includes posting and contouring of contamination data and other numerical surfaces as well as sectioning and fence diagrams through on-screen interaction with the *virtual site*. Figure 5-2 illustrates three graphics developed using Site Planner. The first is a plan view of VOC concentration and aquifer potentiometric surface. The second and third represent contaminant concentrations in multiple aquifers in cross section view, and a fence diagram through a set of wells, also illustrating contamination levels.

An example of a PC application to visualization is illustrated in Figure 5-3, which has been developed using the Stratifact geological analysis software. The graphic includes two typical products: a geophysical bore log related to the stratigraphy and the well construction diagram. The second consists of a cross section through a series of bore logs. This type of visualization is commonly used by geologists and has generally been produced using CAD software. The graphics, in this case, are produced automatically from data stored in the geological database tables, and may require only limited CAD activity to enhance the figure (titles, labels, scaling, etc.).

There are a number of other PC-based software systems such as GIS/Key, GEOKIT, and gINT, which are also PC-based systems focusing on geological data interpretation. Another general purpose visualization system, AVS, which has had applications in medical and manufacturing visualization, runs on engineering workstations, and is also being applied for environmental purposes. Numerous other displays of data in two and three dimensions can be created using these software tools.

The specific features and functions among these packages differ; each has strengths and benefits to their application. The purpose here is not to evaluate or critique these systems; rather, it is to indicate that these, among others, are available to support the analysis and visualization of data. Their results and utility will differ for various reasons:

- Mathematical methods for gridding and contouring will affect numerical surfaces that are modeled
- Graphical software will affect resolution and color range
- On-screen performance for workstations will exceed PC performance; and high-end workstations will out-perform low-end workstations
- Database integration is inherent in some systems, while simple ASCII import of spatial/parameter data is the standard for others
- Flexibility in functions, user steps, and time to produce new views will differ

As federal and state regulatory agencies begin to bring visualization software into litigation, so will regulated industrial and federal organizations need

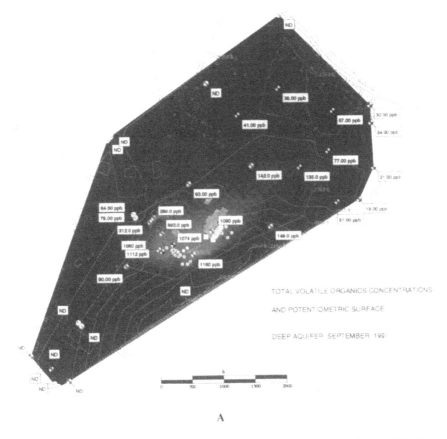

A

FIGURE 5-2. ConSolve site planner virtual site 3 graphics: (A) plan view graphic of VOC concentrations and aquifer potentiometric surface; (B) cross section of contaminant concentrations in multiple aquifers; and (C) fence diagram through wells with contamination

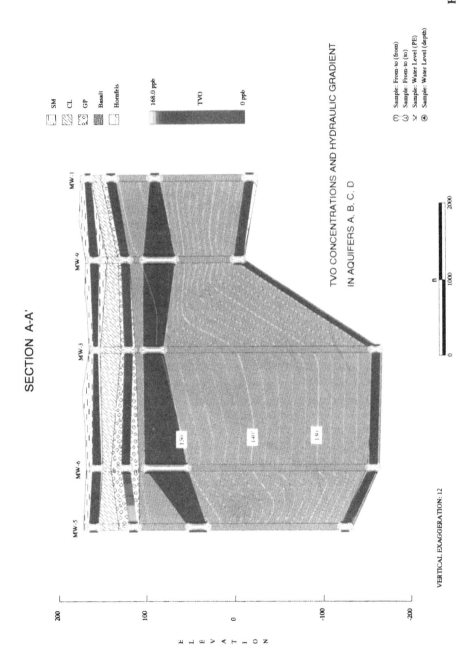

SECTION A-A'

TVO CONCENTRATIONS AND HYDRAULIC GRADIENT
IN AQUIFERS A, B, C, D

FIGURE 5-2B

SECTION A-A'

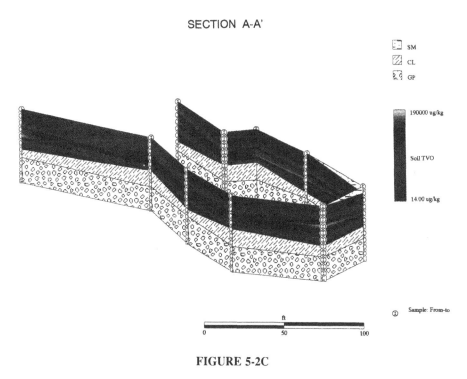

FIGURE 5-2C

to upgrade their applications of these systems. The quality of the visualization graphics is exceptional, particularly in the high-end systems. However, this should not overshadow the scientific quality of the data that support the graphic. In fact, these systems produce results based on models of the subsurface conditions; as such, they are estimates of conditions based on discrete samples taken on the site. Managers need to see these types of graphics to understand conditions and to make decisions, but challenges to the data can discredit even the most brilliantly portrayed graphical visualization. For the visualization to be defensible, it is necessary to understand and describe the nature of the data and the phenomena being explored. When such systems rest on a sound data management process with effective quality assurance procedures to provide accurate and defensible data, they can have significant payoff in presenting a case during litigation.

B. ENVIRONMENTAL FACILITY MANAGEMENT GRAPHICS

The graphics relating to facility management are not always as dramatic in hard copy as those describing contamination plumes. They include both facility features and database tables, and play better on the screen than on paper. However, two examples of facility management graphics are provided below to illustrate the concept.

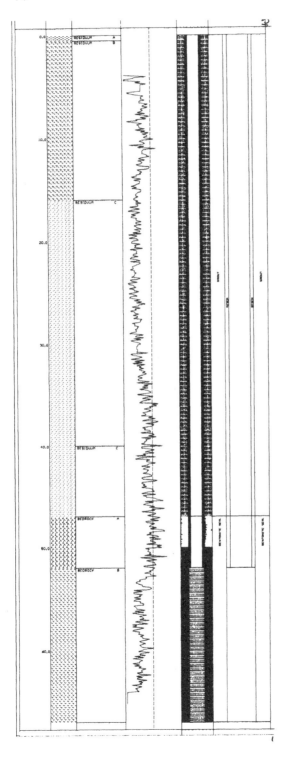

FIGURE 5-3. (A) Stratifact geophysical bore hole log, stratigraphy, well construction, and (B) cross section.

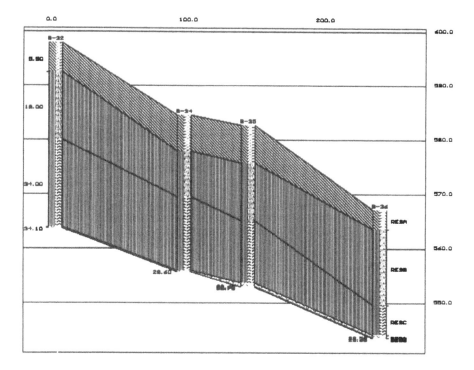

FIGURE 5-3B

The concept of drill-down through database tables is commonly used to access data. A graphical drill-down is a more intuitive approach to this process, since the "driller" is dealing with physical objects illustrated in maps and drawings. This approach is described as follows. Plate 5-6, from *EDGE*, is a screen image of a typical environmental facility management application at XCORP USA, a fictitious company. The map of the U.S. in the top left background illustrates the XCORP facility locations as green squares. A menu provides user functions from which a set of graphical and text windows are opened on the screen through point-and-click and keyboard commands. Proceeding clockwise, the facility drawing for the XCORP Union Plastics plant is illustrated in the upper right corner. This window is activated by pointing at the facility location on the U.S. map. Each element of the map is an object (building, road, waste site, property line, etc.). Pointing at a selected building on the facility drawing, and then selecting the desired floor plan from a pull-down menu displays a drawing of the building's basement floor plan to the right of the site drawing, below the base map (each window is movable to any desired location on the screen). Within the building drawing, another object is evident, which is a process block representing an industrial process (ABC) that is conducted at that location. Using the menu again and pointing at Process ABC object activates the process diagram drawing window in the lower left corner. Process Diagram ABC includes unit operations blocks as well as emission points for air and water. Using the menu and pointing at the Air Emission Point 1 (arrow at the top of the diagram) activates an Oracle Air Emission Database Table illustrated at the top of the graphic.

It is evident from this example that from a map of the U.S. one is able to locate a specific stack emission data table with full Geo-Graphical User Interface functionality. In addition, each of the windows is independently active, i.e., the facility drawing can be zoomed to look at other features; other buildings can be selected and new process drawings activated; other emission points can be selected; and the Oracle data table can be scrolled to view or add data for different time periods or sampling sets and for data editing. Oracle tables could be developed relating to the equipment characteristics in the process, maintenance history, chemicals used, etc., in this same context. Such functionality draws the user into the process of understanding the data and provides a means for enhanced communication of environmental information.

Another example of this sequential drill-down is shown in Plate 5-7, also from *EDGE*. The menus are shown in the background along the left side of the screen. An interactive query using the menus has requested facilities having annual environmental expenditures exceeding a given amount ($1,000,000 in this case). The cost database for facilities is screened and the U.S. map color codes are changed to red. A facility drawing is produced by clicking on one of these facilities. The facility is screened to identify the sites exceeding another input threshold and the site objects turn red (red border around Landfill site). Using the site modeling functions, the chemical database is activated and the contamination contours for volatile organic compounds (VOCs) are produced. These are screened against an input threshold of 500 ppb, producing the graphic in the lower left corner, which also includes the posted values at 6 ft of sampling depth (high spot = 2374 ppb in center of the plume). The Database function is activated from the menu and the complete database describing sampling results for the Landfill site is available in the lower right corner. Specific data for a sampling location are obtained by pointing at the location on the facility drawing; the table is populated with sample results defining the VOC concentrations at that location. The table in the graphic is scrolled to the location LF-08 and the depth of 6 ft and illustrates the same data value (2374 ppb) as in the graphic.

The same type of interaction can produce a bar graph that illustrates the budgeted amounts, actual expenditures, and percent complete (in dollars) for each of the cost accounts in the project control database. This illustrates the linkage between scientific and engineering data (drawings, sample results, contours) and cost data associated with the characterization and remediation process. Facility management issues dealing with compliance and remediation issues and their relationships to costs can be addressed more effectively in this manner.

A third *EDGE* example relates to the drill-down into the image management component associated with the facility. In Plate 5-8, the facility drawing is produced in the same manner as described above as the basis for the spatial queries. A menu of available images is created from the database and is displayed in the lower right corner. Selecting a site location (pointing at the site on the facility drawing) and the desired graphic from the menu activates the imaging software, in this case the Digital Equipment Corporation CDA Viewer. A photograph of the site is selected and displayed (lower center). Using the same approach, a second menu selection for the site produces a menu of the MSDSs available for the facility location (object) that has been selected graphically. A specific MSDS is extracted and illustrated on the left of the graphic. The viewing software can scroll through the pages of the MSDS through the use of GUI commands.

These examples of FM applications represent the future of environmental information management on industrial facilities. They require the integration of available software technologies. Their functionality will differ on workstations and PCs, depending on the machine capabilities and performance capacities and on the specific group of software that is selected (DBMS, GIS, imaging, charting graphics, etc.).

C. EXECUTIVE INFORMATION SYSTEM (EIS)

Another aspect of the above interaction with data and graphics relates to the area of EIS, which are intended to provide access to management data. Plate 5-9, from *EDGE*, is a prototype developed using available Department of Energy data. A graphical display of DOE facility locations is illustrated in the graphic window. A set of object-based, interactive queries have been made against a DOE database obtained from public domain documents. The query menus are shown in the top left of the screen. The color codes of these facility locations on the map depend on the query parameters and relate to costs, management, and geographic variables. The Facility Database Table for a selected facility is illustrated in the lower right corner of the figure. This table is enabled from a menu and by pointing at the facility on the U.S. Map. An Oracle Financial Information Table is illustrated in another window shown in the lower left corner of the screen. This information is available for direct access by a manager for use in planning, technical support to meetings and teleconferences, and for preparation of hard-copy information products in response to requests for information from other managers.

This example illustrates the extension of the environmental FM/GIS approach to management activities. Information is needed at all levels of an organization, and if it is managed and related to facility graphics consistently, its accessibility is facilitated and its utility within the organization is enhanced. FM/GIS technology will become used to a greater extent at executive levels as managers who are comfortable with these technologies migrate to these levels of their organizations.

Chapter 6

SYSTEMS INTEGRATION

Interfacing environmental systems with engineering department drawings and other organizational systems can provide significant corporate benefits. Issues associated with such integration are discussed in this chapter.

The underlying principle behind a GIS-based FM system for ES&H applications is that it is integral to the business operations of the organization rather than a specialized system that is removed from these operations. As such, the system will be enhanced by greater degrees of integration with existing and evolving corporate systems. This requires top managers to understand the values and benefits of integration and to support the policies that will foster movement in that direction.

I. FACTORS AFFECTING INTEGRATION

A. INITIAL CONDITIONS

The starting point for system implementation is with the current operating environment. Initial conditions, as in mathematics with numerical time series approximations, have the greatest impact on the near term behavior of the system. The initial conditions can be defined by the current position, speed, and direction of motion. Organizations have computing infrastructures that have evolved over time to produce a current situation, which may be based on mainframe architecture, personal computer networks, workstations, or wide area networks. The direction of movement may be to a new database, operating system, or GIS platform. There are usually a variety of software tools and applications managed by departments or business units to meet their specialized corporate missions.

B. INFORMATION TECHNOLOGIES

The organizational model based on the production line manufacturing process is being replaced, as is the proliferation of bureaus in government. The winds of change that are blowing in these organizational environments are being felt in the information arena with *open architectures* and *distributed computing* environments. MIS organizations are being driven toward a variety of technologies that offer a better tomorrow, such as *relational database* architecture, which is more efficient and orderly than older technologies, and *client-server* computing in which the databases are distributed over networks and structured for transparency of location to the user. Computer *workstations* and personal computer networks are replacing mainframes and graphical user interfaces are emerging as the means for communicating with information. *Object-based programming, computer-aided software engineering (CASE)*, and *multimedia systems* are being offered as information technologies that afford organizational advantages.

In this environment of change, MIS organizations can sometimes be threatened by service organizations and product vendors offering new technology, or they can be oversold on the advantages of new technology. ES&H managers can become confused over the options and disenchanted with the MIS department for not meeting their needs. Organizational top management can become disturbed over the costs and time frames for migrating to new technology and the undepreciated investment in the current information infrastructure. FM/GIS implementation adds another dimension to this condition.

The implementation of a GIS-based FM approach for organizational ES&H applications requires participation from a number of parties. FM is an engineering function dealing with the organization's real estate, plant buildings, infrastructure, and capital equipment. These same physical elements are the locations where ES&H activities occur. The production line contains the capital equipment that produces emissions and discharges and is the scene of accidents. The organization's Human Resources department has responsibilities for the personnel involved with these processes. The information that is gathered and used flows over communications links and through processors that are the responsibility of the MIS department. This means the implementation of an effective FM/GIS system should be a total organization system in which the ES&H organization participates; otherwise, there will be a proliferation of point solutions and independent technologies without achieving the benefits that can be derived from corporate integration.

In order to take advantage of both existing and emerging technologies for moving toward a GIS-based FM system, organizations must employ three approaches:

- *Systems integration* methodologies, wherein existing resources and infrastructure are connected to achieve the most effective use of available technology for accessing information
- *Structured evolution*, in which many small steps are taken quickly to achieve measurable progress
- *Organizational integration*, wherein the information solutions are dealt with in a total organizational context rather than as simple point solutions for one department

C. INTEGRATING AVAILABLE SYSTEMS

Organizations have specialty skills relating to their own corporate information technology components, but not necessarily in new and emerging technologies, or in other systems that are not commonly available to them. They utilize the services of *systems integrators* to provide better linking among independently applied systems and to implement new technologies and integrate them with their own. There are a number of information technology organizations specializing in the integration of information resources. Product-based information technology organizations that have traditionally dealt in delivering hardware and software in competition with other suppliers are now adopting strategies for integrating with the existing infrastructure, which may be provided by their competitors, rather than attempting to replace it.

Systems integration provides the means for dealing with the current computing infrastructure to bring new technology into use without massive conversion of existing systems. Formalized methodologies can be applied for characterizing an organization and for dealing with the integration steps. In the FM/GIS arena it is necessary to consider items such as:

- Moving drawing and map files from the engineering department and making them useful for managing a facility or for managing the maintenance of plant capital equipment
- Linking human resource databases, safety and health department reporting needs, and engineering drawings to provide means for spatial illustrations of risk or exposure or for complying with training requirements
- Providing standardized electronic access to regulations relating to specific facility locations for the environmental and legal departments
- Installing additional processing and telecommunications hardware and software to facilitate rapid and remote access to site investigation results being developed by an environmental consulting organization
- Linking financial data relating to environmental expenditures from the accounting department to specific waste sites or facilities in order to access graphical displays of cost and schedule status for these locations
- Providing a means for displaying maps and drawings to monitor facility conditions for top management during unplanned incidents or emergency operations
- Bringing together a variety of independently developed tools for site investigations into a system that can support project needs and provide geographic-based access to the information by top or middle managers

While this is not an exhaustive list, it is indicative of the types of *intuitive access to information* and data manipulation that can be made available through GIS technology to support understanding and interpretation of data that reside in a corporate information base.

D. STRUCTURED EVOLUTION

Early attempts at GIS in the 1980s focused on the promise of the technology. These systems were not very sophisticated, particularly in comparison to those that are available today. In many instances, the requirements for acquiring and maintaining effective databases and drawing files were placed in the background. One of the discoveries made during this period was that the costs of implementation were driven by the requirements for developing the drawings and map files and for maintaining currency and consistency in drawings and databases. The cost of digitizing drawings at engineering precision levels was high, the experience base was nonexistent, and technologies were not available to make the process efficient. Implementation of GIS also required significant coordination among organizations having responsibilities for data and drawing components (and still does). In addition, relational database management was not mature at that time. As a consequence of these conditions, there was disenchantment with GIS when the realities of implementation came to the forefront.

To address these concerns, the technology evolved and improvements were introduced to facilitate implementation and use of GIS. These include:

- File exchange protocols and utilities to take advantage of available drawing files
- Automated digitization software
- Better interfaces between relational DBMSs and GIS
- Improved graphical user interfaces
- Increased experience base in use of graphical drawing files

These have led to a proliferation of CAD and GIS applications, and ES&H organizations are faced with a different set of conditions. There is a need now to take advantage of existing systems through a *structured evolution* approach. The objective is not to create an ES&H system, but to evolve it around the available resources within the organization, not as a point solution but as integral with the organization's operations. The important steps to be taken are:

- Create a *data model* that meets the ES&H needs
- Identify the locations in the organization where relevant data is stored and can be used to support the data model
- Determine which areas of the organization will benefit from automation
- Develop methods and procedures for moving data and drawings into the ES&H system
- Modify and refine the procedures to improve the efficiency and performance through continued improvement of the system

The first attempts will be faced with choices: more time and costs to develop efficiency or immediate results using available data and systems. The

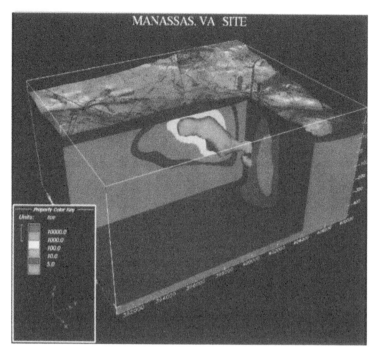

Plate 5-1. Earthvision chair cut visualization of sub-surface soil contamination.

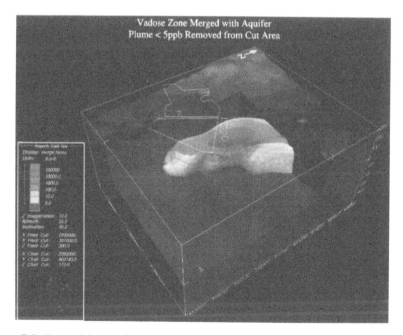

Plate 5-2. Earthvision chair cut through the vadose zone and aquifer with surface features in transparency.

Plate 5-3. FMC visualization of contamination plume beneath gridded topographic surface including wells and surface features.

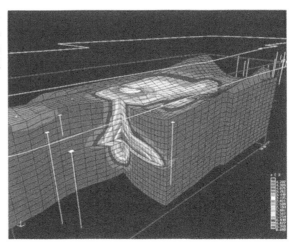

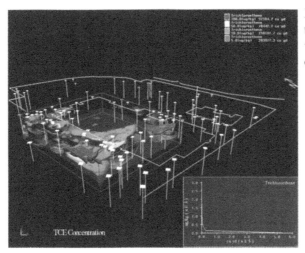

Plate 5-4. FMC visualization of surface features, wells, and sub-surface contamination.

Plate 5-5. Intergraph voxel visualization.

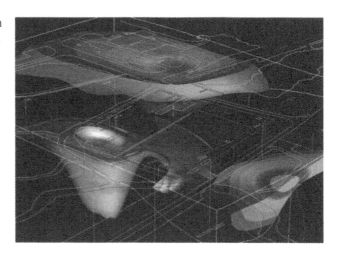

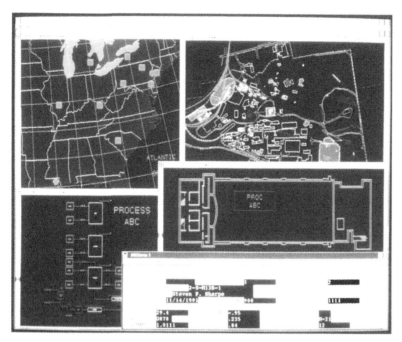

Plate 5-6. *EDGE* screen FM image of geo-GUI drill-down to facility, building, process, and air emissions database.

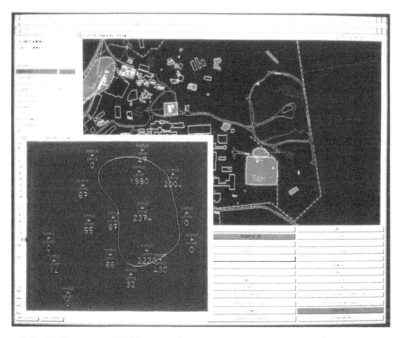

Plate 5-7. *EDGE* screen FM image of open window for site drawing, contamination zone, and chemical concentration database table.

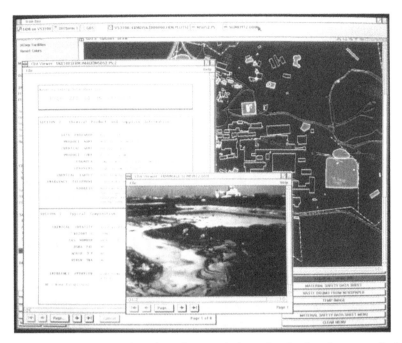

Plate 5-8. *EDGE* screen FM imaging of open windows for site drawing, remediation site photo, material safety data sheet for site, and image selection menu.

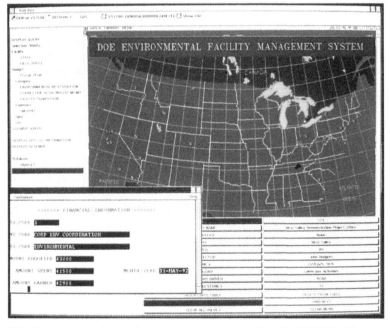

Plate 5-9. Executive information system screen graphic for DOE facilities using *EDGE* approach.

latter approach may be undesirable from the view of a system designer but will be acceptable to the ES&H manager's immediate concerns. It should be understood that improvements can be achieved once the feasibility and utility of the approach have been demonstrated.

The first time an existing data set or drawing file is obtained and used in the ES&H system, it will be necessary to invest resources in moving them from their natural locations and formats into the required structure. The file transfers may be accomplished through the preparation of specialized files in ASCII format or in database format, depending on the organization's operating environment. Then it will be necessary to implement modifications to comply with the data model, remembering that the data were there before the data model, and they may not fit together perfectly. The modifications may be with codes, field lengths, formats, naming conventions, or other factors that will become standardized within the new data model.

As the process is being carried out for the first time, it should be documented, so that it can be repeated readily. Subsequent conversions will be easier, and methods can be developed to make them more efficient; e.g., network transfer might replace the original diskette procedure; the modification to the files might be automated, replacing the manual procedure; or the source files might be modified by the owners prior to delivery to the ES&H system. Pilot projects can be employed to conduct these improvements. Progress can be evaluated by measuring the time and effort required to accomplish each subsequent file transfer. MIS analysts involved in the evolution may identify approaches for achieving large-scale savings, and these should likewise be subjects for additional pilot studies.

The same approach can be applied to drawing files. Initially, a pilot study should be conducted, using a set of facility or process drawings based on their available format and using conversion steps to load them to the GIS. They will probably lack the drawing intelligence that is needed, and this will have to be added afterwards, within the ES&H system. The pilot study should demonstrate the feasibility of using these drawings and the approach for making them useful in the ES&H context. Once this has been accomplished, the decision might be to complete the same steps for several facilities. An alternative decision could be to consider modifications of existing standards within the engineering department and to apply these to subsequent drawings to meet the ES&H need. Perhaps a set of new facility drawings is contemplated in the near term and modifications to standards can be accomplished with little difficulty. There is a need for understanding and compromise for the organizational whole in this regard.

E. DRIVERS

U.S. organizations have pursued the *total quality management* philosophies of the 1980s in varying degrees to improve their operations and respond to global competition. They are now embracing the *re-engineering* of processes to achieve productivity and economic benefits in the 1990s. Integration of information resources is a manifestation of the re-engineering philosophy, in that it seeks to organize and utilize available resources more effectively. The orderly flow of information within a large organization is not easily achieved. An organization that makes strides in this area will be more competitive and responsive to the regulatory and economic forces affecting its competitive status and position in the market place.

As ES&H activities assume greater levels of visibility and costs, executives must devise approaches for managing and reducing these costs. ES&H organizations are cost centers that may consume as much as 3 or 4% of an organization's revenues. They deal in information rather than marketable products. Efficiencies achieved through re-engineering the ES&H process can have significant impact on financial performance and can increase stockholder confidence as well as public image.

II. THE PLAYERS

There are many parties involved in the implementation of a corporate ES&H system. Additional parties become involved when it takes on GIS features. These are *the players* in the process. They may pose questions and objections such as:

- Why do we need their data?
- How will we get all of this data in?
- Who is going to maintain it?
- Why can't we just use the drawings we have?
- Where do we start?
- Do we really need all of that?
- How are we going to pay for it?

These issues are real ones and cannot be dismissed without good answers. They require homework and must be understood by both MIS and ES&H organizations and can only be addressed effectively if the top management of the organization is supportive of the process.

Each of the organizational components raising these issues has a mission that will be affected by the system. The implementation will require better information exchanges among the variety of corporate systems where data reside to support the ES&H activity, which will use information from several organizations, for example:

- Engineering — plant and process drawings and equipment specifications
- Production/operations — product quantities and production rates
- Maintenance — maintenance records and waste disposal records

- Human resources — personnel labor histories, medical records, and training status
- Legal — regulations and litigation status
- Environmental — permits, emissions, site investigations, schedules, and policy documents
- Public relations — policies, incidents, and media coverage
- Safety and health — permits, incidents, emergency plans, inspections, training schedules, and policy documents
- Shipping and transportation — deliveries and material descriptions
- Accounting — ES&H incurred costs and production costs
- Finance — ES&H budgets, work breakdown, and schedules
- Information resources — data flows, quantities, locations, structures, and available systems

It is important to recognize that these information requirements relate to effective management of ES&H regulatory compliance issues. Clearly, the other participating organizations have their own missions and associated roles to deal with, and their own information requirements. The objective is to obtain the appropriate information from them, use it for ES&H purposes in the plant, and make it accessible to managers for planning and decision-making purposes.

A. ORGANIZATIONAL INTEGRATION

It becomes evident that the structured evolution approach has an inherent *organizational integration* component. Information decisions have a way of making organizations face their structural shortcomings. Information is data that have been lifted to a higher state (processed, organized, and in context). Much like the relationship between energy and entropy, the data model provides order to the process and increases the information content and the value of the data. The process provides with it the added benefit of organizational integration.

Over time, drawings and data will change and a more efficient system can be evolved as the complete drawing inventory is modified. In planning the evolution of the system, it will be advantageous to define specific steps that will require investment and to evaluate the expected benefits from these steps. MIS, engineering, ES&H, and human resources budgets might be pooled to develop an integrated project that bridges the department needs, providing leverage for achieving a greater overall benefit — another organizational integration step. Staff from each participating group should understand clearly the objectives of the ES&H implementation strategy and identify methods to add value that will advance their individual needs. By identifying and evaluating these openly, compromise positions can be reached, benefits will be perceived, and the evolution can proceed.

III. THE INFORMATION

Knowledge is power, and it is derived from the information that flows in a complex industrial organization. Much of the information that drives the ES&H system is not unique to that organization's activities; rather, it is information that is inherent to the operations of the organization and is maintained by others. There are some exceptions, for example, specific site investigations or testing programs that are managed by ES&H departments. Integration with existing systems is an essential component of the implementation process.

One cannot walk down the halls of an office building without seeing countless computer screens employed in accessing, displaying, creating, and reporting information. The information is constantly flowing at various rates and quantities and at different levels of efficiency. Resistance to flow is encountered when interfaces among sources and needed outputs are inefficient. This can occur for a variety of reasons. Reducing this resistance is one of the essential missions of the MIS component of the organization.

The system planner must understand both the information that is needed to meet the ES&H requirements and that which is available throughout the organization. This can be a formidable task requiring assistance from many of the organizational component groups. One can expect that a variety of electronic data sources and formats will be identified when characterizing these sources of information. The process of understanding what is available can be achieved by developing a summary table that includes the following elements:

- Type of information
- Responsible organizational component
- Responsible person(s), location(s)
- Information system identity, location
- Software tools used by the system
- Format, structure, and quantity of information
- Currency and update frequency
- Users of the system, and locations
- Information products available from the system
- ES&H application of the information from the system

This summary provides an overview of the information that is available, e.g., data, maps, drawings, documents, images, figures, reports. It also identifies how that information applies to the needs of the ES&H system.

The findings of this survey can be both exciting, in the discovery of valuable information assets, as well as discouraging, because of the perceived effort for making it work together in an integrated manner. However, the flow of information is always difficult to achieve in comparison to designing a database structure, and one should not be too discouraged. Fitting the pieces together is the next challenge.

IV. THE DATA MODEL

The term *data model* is used commonly in the information systems world and in GIS development. Here, it is defined as the structured organization and flow of information, which identifies the data categories and their components in a manner that relates to the operations of the organization. There are formalized procedures and software (CASE tools) for developing data models. These tools assist in defining the data structure based on defined elements and activities of the organization, called *entities*. Some of them can also produce data tables and software. These tools are useful in helping MIS staff to describe the requirements in logical terms and in having the resulting models reviewed by ES&H staff for consistency with requirements and objectives.

A. PURPOSE

The data model for the system should address the ES&H organization's needs for managing the information types that are required to achieve compliance with regulations, support planning and strategy development, and measure performance of ES&H measures against strategic goals. It should be capable of addressing the full spectrum of needs from shop floor issues dealing with reporting hazardous chemical inventories and usage to executive-level reporting of trends in incident occurrences or emissions per unit of production. It should describe the ideal, unconstrained structure and flow first, before it becomes constrained by the realities that are identified by surveying the availability of information and by the difficulties that will be encountered in achieving efficient information flow within the existing systems.

The freedom to design a system on a clean blackboard (or whiteboard) is usually not available to information system planners. There are usually constraints to the design that are caused by the capabilities and structure of those systems that are in place, which have been designed and implemented at an earlier time for their own purposes. Their data elements and graphical components should not be expected to fit readily into the ideal model.

Integrating these systems can be achieved most acceptably through progressive evolution rather than by radical change. In the existing state there may be no flow paths available to move information where it is needed, or perhaps there may be limited flow capacities. There may also be redundancies and other inefficiencies from the standpoint of the ES&H system requirements, since the information has not been intended for these purposes. However, the utility of the available information can be enhanced through ES&H use, and benefits can be derived by the organization through the creation of information flow where there was none. The extent to which information flow will be increased depends on the organization's management structure and on its ES&H policies and procedures.

B. ELEMENTS OF DATABASE STRUCTURE

The database structure can be defined in terms of *tables* that describe logical entities or groups of data, and *fields*, which are the data elements stored in the tables. Each table contains a set of fields; and the same field may appear in more than one table. The relationships among tables are defined by selected *key fields*, which are common to two or more tables and provide the relational context of a relational database structure. These logical structures are essential to the effective design of the database. The efficiency, utility, flexibility, and speed in providing users with information and in reducing the overall size and redundancy of the database are all affected by the database structure.

The database structure can be described by *entity diagrams* that convey the organization of and the relationships among its tables. The data elements within each table are defined in terms of their database field name, description, and format. A *data dictionary* identifies and describes each field in the database and the tables in which it is stored. Implementation of the database within a specific DBMS tool can be accomplished readily from the entity model and the data dictionary.

The entity diagrams for the following databases are described below:

- Emissions and discharges
- Site investigation
- Executive information system for site prioritization

Each database has been designed in a GIS context, in which the relational database has been linked to CAD drawings containing the same location codes as those in the database.

C. EMISSIONS AND DISCHARGES

Figure 6-1 illustrates a database structure that is applicable to the management of emissions and discharges from an industrial facility. The structure is illustrated in a hierarchical format, although the model has been implemented in a relational database management system (RDBMS). The tables characterize the industrial processes that produce the emissions. Each table contains appropriate fields to describe the table entity. A pair of tables are linked by key fields, which are common to them. The hierarchy implies levels in the structure. The lower level *(child)* tables cannot be completed without higher level *(parent)* records. For example, there can be no Process Line data until a Facility table is defined, and data for Equipment in the plant cannot be entered until there is a Process Line to which the equipment can be associated. The model permits arbitrary numbers of Facilities, Process Lines, Emission Sources, etc.

89

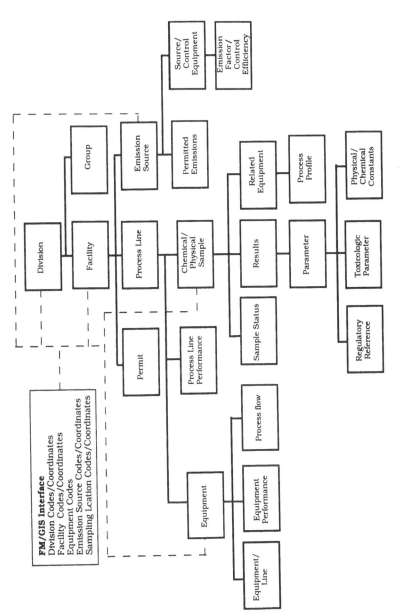

FIGURE 6-1. Emission/discharge database entity parent/child diagram.

The table structure is capable of including data for air, waste water, and solid waste emissions through the structure of the Emission Source Table. Likewise, the Control Equipment can include scrubber, baghouses, and other air control equipment or waste water treatment equipment. Equipment is associated with one or more Process Lines. The Process Line Performance Table defines the line throughput parameters and values which may be qualified as averaged, measured, specified, permitted, etc. The Process Flow Table describes how the Equipment is connected from a process standpoint.

The measurement of emissions is accomplished through stack or waste water sampling. The Sample is described in the Chemical Physical Sample Table and the laboratory analytical results are stored in the Results Table. The Parameters and regulatory references to these parameters are also included in the table structure. The Permits are associated with the Facility, but may relate to a specific Line or Equipment through relational references. Emission Sources (stacks, outfalls) are coded along with their Permitted Emissions. Emission Factors and Control Efficiencies associated with Source Equipment and Control Equipment are related to the particular equipment in the database.

The complete data structure for air emissions stack testing was developed to include the laboratory analytical results for these tests. The database was designed to incorporate wastewater and solid waste discharges directly as an evolutionary step.

Another important aspect of the data model relates to its implementation. Two specific requirements were defined and included:

- The laboratory data structure was designed to emulate the laboratory procedures for submitting test results on diskettes. Laboratory results are batch-loaded quickly into the database as soon as they are delivered. The appropriate sampling event codes are included on these diskette submissions and must be consistent with the sampling events required by the database tables.
- Process flow diagrams for the Process Lines in the Facility are prepared first on a CAD drawing to include the appropriate codes for equipment identities, emission points, and sampling locations. The CAD files are processed to extract these data elements and batch-load them into the appropriate data tables to assure consistency between drawings and database fields, achieve efficiency in data loading, and reduce errors. This approach facilitates future evolution to full GIS/FM implementation.

This design and implementation approach illustrates the principles of:

- Structured evolution starting with air emissions and proceeding to include all plant discharges
- Integration of existing systems for delivery of laboratory analytical results and rapid loading into the database
- Integration with the engineering plant drawing system

D. SITE INVESTIGATION

A relational database structure used to manage data for site investigations is illustrated in Figure 6-2. The parent-child hierarchy is illustrated in the figure. The organization is divided into Divisions, for which there are Facilities (plants) at the top of the structure. The Facilities can be organized by Group (common remediation problem, subcontractor, etc.), and a Facility may be divided into Areas (waste sites, operable units) or Zones (Areas or portions of Areas). The Location (x,y,z) where samples are taken is a critical component of the data structure, since data elements in the site investigation tables consist of field data, and laboratory analysis results must be related to physical locations on the site. The remaining tables contain the site data that are to be included in the project database.

The structure is also designed for integration with FM/GIS applications; in particular:

- The Division and Facility coordinates may be used to locate all of the organization's divisions and facilities on a single base map
- Location codes and their x,y,z coordinates are used to display chemical concentrations; they are also related to drawing object codes for facilities, buildings, emission points, and waste sites to provide geo-GUI access to data tables, and to relate cost data and images to these objects
- Depths and depth intervals are used in conjunction with location codes to produce stratigraphy and ground water surfaces that can be viewed in cross section format or in three-dimensional perspective within the FM/GIS

Batch loading of soil and water sample laboratory data is achieved through coordination with contractor laboratories in the same manner as with air emissions results in the prior example.

E. EXECUTIVE INFORMATION SYSTEM FOR SITE REMEDIATION AND PRIORITIZATION

A third example of a database structure is provided in Figure 6-3, which describes an executive information system (EIS) application. In this case, the organizational structure included the Group and Division Levels. The Site was used to designate the facility, since it was a commonly used and understood term. In this structure, a Waste Site is called an Operable Unit (OU), which is commonly used in federal site remediation programs. The purpose of this system is to analyze and compare sites and OUs with regard to their priority, status, and their expected costs of remediation.

The data structure includes a description of each Site Stage (of the regulatory/remediation process), the Cost Recovery method to be used, and a Prioritization Table that includes both objective and group judgmental variables

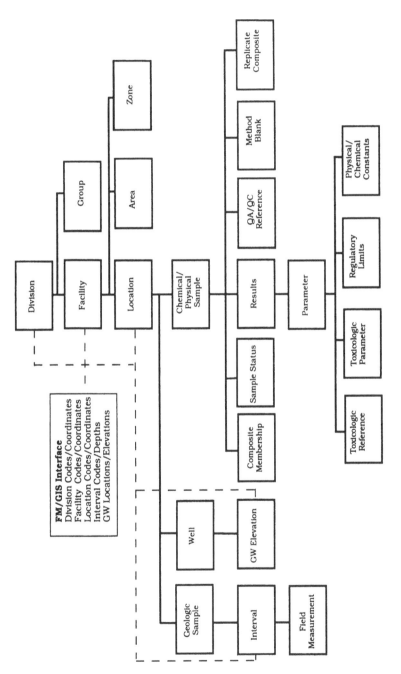

FIGURE 6-2. Site Investigation database structure.

93

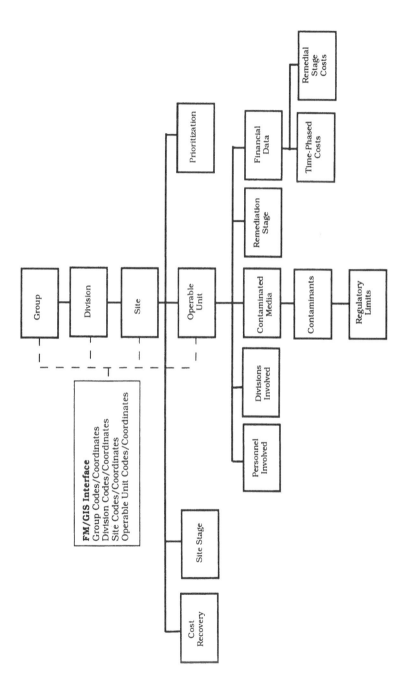

FIGURE 6-3. Executive information site remediation database entity parent/child diagram.

relating to public perception, expected costs, potential liability, local presence, etc. Below the OU Table are other tables describing the Personnel and Divisions associated with the Site/OU, the nature of the Contaminated Media, and Remediation Stage and associated Financial Data.

Typical steps required in implementing this data structure include:

- Use of existing scoring systems or developing new algorithms for prioritizing waste sites
- Converting hard copy reports of facility and waste site status into database tables
- Use of corporate financial information
- Developing relationships of data tables to map objects (Sites/OUs, etc.)

In this manner, the system is capable of developing map displays of facility location GIS queries to produce specialized map/graphic products and on-screen tables relating to scores for facilities and waste sites.

These requirements illustrate the integration principles that were described earlier. The flow of data into the system included:

- Extraction and coding of data from existing hard-copy documents into spreadsheets
- Extraction of data from existing organizational databases
- Batch loading of spreadsheet data into the relational DBMS
- Approximate spatial location of facilities on U.S. and individual state maps

The resulting implementation was in the form of a pilot project that demonstrated access to the data and provided hard-copy graphics describing sites geographically, including color-coded displays to describe:

- Business unit identity
- Multidimensional cleanup priority scoring levels

Multiple queries can be overlaid on the same graphic display to further illustrate spatial location of high-scoring sites by a business unit. The system could also be applied to strategic planning; for example, developing average distances from waste sites to selected locations, with the objective of developing a corporate organizational structure that includes preferred geographic locations where management of waste sites would be conducted within each business area.

F. INFORMATION PRODUCTS

System designers usually emphasize the data model and database design, while users are more interested in the information products that will be produced by the system. The designer must be concerned with satisfying the user's necessary requirements, which may be to produce compliance reports through automated methods. A sound data model will facilitate the development of these reports and will also provide the flexibility to create other valuable information products that may not have been considered by users. It should, in essence, anticipate the questions that might be asked of the database by managers and planners. It should assist in defining the appropriate technologies and not misusing technologies (raster-based GIS for discrete feature manipulation, or layer-based system when object identities are essential).

A large percentage of system users may have primary needs in maintaining data for regulatory compliance reporting, while managers may have need for ad hoc reports and summary information. Often these management information products are not specified clearly, but a good analyst who understands both the database structure and the management need can produce valuable ad hoc reports readily. The types and variety of information products that are available when an integrated data model is available to the organization are limitless. When a pilot project is used to evaluate data models and prototypes for system integration, there should be an appropriate balance of emphasis between standardized reporting and ad hoc information products.

V. REGULATORY INFORMATION

The laws and regulations that drive the ES&H requirements represent a public information source that can be integrated into the system, not spatial data. Regulations are usually maintained in hard copy or electronic format within an organization's legal department. They are also used by ES&H managers and plant level personnel to address reporting needs and to deal with operational issues.

A. FEDERAL AND STATE REGULATIONS

Electronically formatted regulatory information can be purchased from firms specializing in ES&H regulatory publications. The information may be available in magnetic storage media on diskettes or tapes or in optical media on CD-ROM. The offerings are accompanied by licensed software to access the information through menu queries for key words or regulatory titles. Updates to provide current documents are made available at regular intervals in conjunction with the terms of the purchase or license. Information in such systems usually includes federal and state regulations. Full regulations or abstracts may be used, or selected regulations may be acquired that deal with the specific activities that are being conducted.

B. LOCAL AND SPECIAL REGULATIONS

Local regulations below the state level are generally not available electronically in the public domain. An organization may decide to build a customized electronic file for this purpose. In addition, specific legal requirements may be imposed on the organization, such as consent agreements, notifications, and fines. These can be handled by document management systems that can be used to manage access to document images. A database indexing interface is used to identify document titles and subtitles, which are stored in a database. These titles are indexed to pointers that locate them within the image files. The database storage component of these systems is small in comparison to the number of characters stored as graphic images, but it is an essential one. Document navigation can be achieved quickly and efficiently through a well-indexed database. This approach can provide a means for integrating electronic access to regulations that are not available commercially.

C. INTEGRATION ISSUES

Access to the regulations may be over a local area network or wide area network, on a mainframe, or on individual personal computers. Since federal regulations have a broader need than state and local regulations, it may be desirable to use a network or mainframe for these and PCs for state regulations. An appropriate design electronic regulation access will depend on the available hardware resources and network availability.

Integration of electronic regulations within the system structure must deal with the following issues:

- Plants or other organizational entities may already subscribe to an electronic publisher of ES&H regulatory information
- Existing document management systems within the organization may be available to facilitate the implementation
- Customized indexing of information to locations requires organizational standards for achieving consistency and skills and intelligence in those performing the indexing
- Regulations must be linked to facility locations through the use of compatible location codes that can be related to title indexes

Researching regulations can be a tedious process, so that reducing the time and effort required to find the regulation and extract the required information is a worthwhile undertaking. The data structure can be used to identify those regulations that apply to each facility. Additional details and more specificity may be included by relating particular regulation subtitles to discrete buildings or sites on the facilities. The extent to which these steps are undertaken will depend on organizational issues and associated costs and benefits and should be investigated before selecting an approach and embarking on an implementation path. Technologies are available to support the integration of regulatory data within the ES&H system; many benefits can be derived by applying them. The structured evolution approach can be applied by first, defining and understanding the information flows, and then identifying meaningful pilot projects that can demonstrate the feasibility and utility of integrating this information component.

1. PROTOTYPES AND PILOT PROJECTS

Implementation of an ES&H system that relies on a project requiring a long-term development cycle will not be acceptable to most organizations. To address the requirement for shortening the implementation cycle, the *rapid prototyping* approach is an effective methodology for development of new systems. The advances in software development tools facilitate the creation of prototypes that provide users with the *look and feel* of the ultimate product and also create the appropriate user interface. The reports and graphical information products that will be produced are linked to the interface to establish a firm basis for interpreting user requirements and for assuring users that these requirements are being interpreted correctly. The prototype should be available within a short period of time, ranging from a few weeks to perhaps 1 or 2 months, depending on the scope of the system and the extent to which the prototype performs actual or simulated functionality.

Once the prototype has been developed, it should be demonstrated for users. Demonstrations can include:

• Interface with the database through data entry and editing screens
• Interface with drawings and links between drawings and data
• Preparation of information products: reports, maps, and graphics
• Ad hoc use of the system

The purpose is to obtain comments and feedback concerning the system's expected behavior and performance. The comments may include substantive issues having significant effect on the data model as well as cosmetic changes to screens, reports, and terminology. Prototyping enhances the communication between system developers and the user community, with the prototype serving as the focus for the dialogue. User reactions are provided early in the development cycle, before significant investments in full system development have been made, so that there is time to have an impact on the system structure and functionality. User comments should be recorded and then categorized and documented. Another evaluation of the prototype can be scheduled after these comments have been addressed. The same approach for recording and acting on comments is appropriate, and the users and developers should be achieving closure in communicating requirements and functionality at this point.

The prototype provides a firm basis for determining the time and effort required to conduct a *pilot project* with an operational system, perhaps in a single-user mode, or on a single machine accessible to several users. The purpose of a pilot project is to test the utility of the operational prototype. A structured evaluation procedure is useful in defining the value of the prototype system and assists in identifying shortcomings in an operational environmental and enhancements that can improve the system's utility. Objectives for the project should be defined and performance measures identified to evaluate the project. The selection of plants and specific users for the pilot project are key decisions. The severity of the test will be determined by these considerations. Although some believe that a severe test should be given at an early stage, it is usually preferable to define a pilot project that has a high likelihood for success. While a set of conditions for failure can generally be defined, the objective should be to achieve incremental and continuing successes, justifying the decision to develop the system, and the resources being directed to the effort.

The elements of a successful pilot project are:

- Strong advocates for success within the development group and at the user sites
- A manager who understands the regulations and the information technology requirements
- Organizational commitment to funding and proceeding with subsequent steps
- Clear objectives, test plan, and performance measures
- A documented evaluation of results, findings, and lessons learned

After a single-user pilot project migration to a multi-user network, perhaps several plants or a single division of the organization can follow. The same approach for defining objectives and evaluating success apply. Adoption as a corporate-wide standard system would follow, assuming favorable evaluation of the multi-user system.

The points to be made in this implementation strategy are:

- There are many ways to satisfy a requirement
- The optimal approach requires an assessment of the life cycle of the application, not simply the purchase price or development costs of the software
- The long-term support of the system, either through in-house staff or through licensed or contracted maintenance, are essential components of the evaluation process

- Flexibility of vendor offerings that afford integration with in-house systems and customization can provide important advantages to an organization
- Evolution through prototypes and pilot projects provides valuable information and learning as well as an opportunity to adapt to changing regulations and technologies that are more than likely to occur
- Costs are incurred systematically in increasing amounts based on measurable progress
- Hybrid approaches and new integration ideas are likely to arise on the basis of experience gained, which can further improve the overall solution

Every organization has general requirements for compliance as well as special needs that derive from its structure, position in the market, and economic conditions. Organizations should learn from the experience of others, but should adapt these available solutions to their specific needs. Organizations interested in a FM/GIS approach to ES&H information management should consider these issues and the structured evolution approach for system development in achieving their objectives in this area.

Chapter 7

SYSTEM MANAGEMENT

This chapter describes the system management of computer hardware and software that is appropriate to meeting the requirements of effective environmental facility management. The application of personal computers, workstations, mini-computers, and mainframe processors is described. Software requirements for meeting the needs of environmental FM/GIS applications are discussed. Accessibility of the system and the information products it produces to the organization via telecommunications are discussed in this section.

A chapter about hardware and software for FM/GIS applications is somewhat like a newspaper; it is obsolete tomorrow. The dynamics of information technology advances in the 1990s are dramatic; there are daily announcements of new system capabilities and performance. Order of magnitude changes in processing speeds and storage capacities are routinely achieved every few years, and costs per unit of capacity continue to decrease. Corporations that were the historic leaders of the industry are facing financial crises, and they are being forced to rethink their position in the market. Many of the shining star organizations that have emerged in this new era are already history; others have stalled, are regrouping, or have been replaced by a new wave of challengers.

These conditions raise doubts in the market as to which of these players will succeed. Where should the bet be placed when an organization is making decisions to acquire new hardware and software and to implement new systems? The reluctance to commit to a specific technology is understandable under these conditions. The objective here is not to select the specific information technologies and applications, but to discuss in broader terms some of the considerations in this area.

I. COMPUTER HARDWARE

Personal computers (PCs) have become common to virtually all business and scientific applications in both industry and government. As the capabilities of these machines have grown, they have taken on the roles that were relegated to the large mainframe computers of the 1965 to 1985 era. They have more power in a smaller package and are usable by everyone. They are now marketed as commodities, because standards have made them interchangeable; like commodities, their prices are volatile and competitive. Most requirements for new systems are being defined in terms of PCs and networks.

A. PC AND MINICOMPUTER WORKSTATIONS

During the 1980s, the engineering workstation, a minicomputer designed for personal use in scientific and engineering applications, and UNIX operating systems emerged in the market. The power of a workstation in the hands of a single, interactive user fueled major growth in GIS applications where the processing of large map files was required. Workstations provided the power to deal with complex, technical computations and produce graphical displays of information within short time periods, acceptable to interactive users. Most of the currently available FM/GIS systems and applications grew out of mainframe and minicomputer applications and have now migrated to engineering workstations and personal computers.

Engineering workstations using minicomputer ideas and operating systems are configured differently from the DOS PCs that have dominated the installed base for conventional business applications. They were designed to be more effective in FM/GIS applications, both to process data and to transfer it quickly into screen images.

PC applications to FM/GIS in the early and mid-1980s were limited. Their performance was below that of workstations, and below the desired levels for most true FM/GIS applications. This trend has continued, and new generations of workstations are being introduced to preserve this margin. However, the PC evolution has provided machines having capacities that are more than acceptable for FM/GIS applications. While the evolution of workstations has provided a performance margin, particularly for large graphics files and sophisticated FM/GIS applications, the differences between these two hardware types are becoming clouded. The emergence of the Windows operating system for PCs has been a major factor in this area. These differences are now in the chip architectures, as they affect processing speed and data transfer rates, and in the compatibilities of running software applications. The PC is a desktop unit that may be used as a personal machine, as a file server, or as a network controller. Workstation applications, because their genesis is from minicomputer architecture, have portability to multiuser applications that are generally not available in PC configurations.

It is important to recognize that the ES&H system is integral to the organization's operations, and that the users are not a homogeneous group. The capacities of PCs vary significantly in an organization, depending on the software that is installed and the abilities of the users. This will be true in the FM/GIS environment of the ES&H system. Engineering applications, where three-dimensional modeling, vector processing, and high-speed graphics are essential, may find a workstation environment to be advantageous. For *view-only* access to information and for less intensive applications, PCs can be a less costly choice. However, these statements must be taken in the proper context. The appropriate configurations within an organization may include combinations of workstations and PCs, networked in a manner to optimize the power needed by each of the user groups. An integrated system that takes advantage of the organization's hardware and network architecture and user requirements can derive benefits from both workstations and PCs.

B. MAINFRAMES AND MINICOMPUTERS

Corporate mainframe computers are not suited to FM/GIS applications. Mainframe systems are designed to handle large numbers of users in support of a variety of software applications. These multipurpose systems cannot be taxed with an FM/GIS application requiring large graphics file transfers at interactive response times. Their response performance is decreased dramatically with each GIS user. This requirement will cause the system to become congested and the user response time will not be acceptable. FM/GIS applications perform best in a dedicated operating environment, independent of the other broad range of corporate applications.

A dedicated mainframe or minicomputer could be used for an FM/GIS application. Multiuser FM/GIS applications are used in dedicated minicomputer environments, where this application is the sole purpose of the system, and perhaps for economic reasons is based on corporate computing strategies that are in-place. The user interface is similar to the dedicated workstation environment, except that access to the FM/GIS software and data files is through unintelligent terminals or through PCs configured as terminals. Minicomputers are capable of handling large numbers of FM/GIS users with acceptable response times. PC access to an FM/GIS minicomputer via terminal emulation offers the advantage of dual application to the PC: as a workstation in an FM/GIS environment and as a stand-alone machine or in LAN operation for other purposes.

Finally, mainframes and minicomputers having multiple users are centralized systems. The costs of GIS and DBMS software are much higher for these systems than for single-user PCs or workstations. Furthermore, an existing piece of hardware that was not intended for FM/GIS applications is likely to need upgrades in memory, I/O capacity, disk storage, and graphics terminals to achieve acceptable performance. The use of such a machine can serve as a transition point. A pilot project might be used to install and evaluate the application in a multiuser environment and then evolve to a more modern workstation/PC networked system.

C. MEMORY AND DISK STORAGE

The demand for conventional applications such as word processing, spread-sheets, and databases has resulted in an extensive variety of choices and performance improvements for the PC user community. Users have become accustomed to instant response performance for character-based applications and to similar quick response for the newer GUI applications.

FM/GIS applications may require continuing access to large numbers of drawings, which must be stored and retrieved for display on the monitor screen. PCs were not designed to manipulate graphics efficiently at high speeds, as were engineering workstations. Traditional PC users who begin to experiment with GIS applications are often surprised by the much greater response times that are experienced in these applications. Loading FM/GIS software onto a PC that has been effective in traditional nongraphic applications is likely to result in unacceptable performance. PCs having graphics capabilities required for large FM/GIS applications are becoming available, but managers must understand that the basic machines used for word processing and office automation purposes are not appropriate to these applications.

PCs or workstations used for FM/GIS applications should be configured appropriately. The configuration relates to the processor speed, which is commonly understood to affect the rate at which instructions are processed; however, for GIS applications, the memory and I/O (input/output) transfer rates are equally important. The processing speed and I/O capacity are inherent to the PC or workstation architecture. Moving a drawing from a magnetic storage medium to a screen display requires the transfer of large files, perhaps several megabytes in size, to produce the original drawing as a screen image. The drawing speed will be affected by the processor capacity, bus speed, graphics card, and memory available for graphics.

Refreshing the drawing based on user commands occurs often during a user session. If the drawing is available in memory and the I/O capacity of the PC is high, the redraw can occur quickly. If the system must retrieve the drawing from the disk where it is stored, undesirable delays will be encountered during redraw. PC memory costs are low, and investment in additional memory is essential for these applications. "You can never have too much memory," is an adage that applies in this arena. This holds true for workstations, as well.

Because of the sizes of graphical drawing files, disk storage requirements for FM/GIS systems are higher than for conventional character-based applications. Disk storage capacities of several thousands of megabytes (gigabytes) are commonly required for systems where a significant number of drawings are being maintained for immediate access. PC disk drives of several hundred megabytes or in the gigabyte capacity range are commonly available. Disk storage should be purchased in abundance for FM/GIS applications.

D. MONITORS AND DISPLAYS

An important hardware consideration is the monitor and its graphics adapter. FM/GIS is inherently a graphical environment and the visual images confronting the user are an essential component of its acceptability. These are affected by the size of the screen and its resolution. Color monitors are essential, and screen sizes larger than 19 in. are advisable for most applications, although smaller sizes can be appropriate to casual users for viewing graphical information produced by others for review and editing. Higher resolution screens provide more *pixels* per square inch, so that lines have an appearance of continuity, rather than the granularity of dots at lower resolutions.

Large-screen, high-resolution monitors can often be more expensive than the PC processor used for the application. These costs are generally offset by its improved human factors and ergonomics when it is to be utilized regularly and for long periods.

When group viewing is required, large television screens can be used. Some television monitors are configured for Video Graphics Array (VGA) signal display, and there are hardware devices available to convert VGA signals for video display. In addition to monitors, projection systems are available to display screen images. PC systems display the VGA monitor signal output to an overhead projector. Three-gun projectors are also used for VGA and Analog Red, Green, Blue (RGB) output. As the information highway technologies evolve to link computer, television, and telephone, one can expect further developments in this area.

E. DIGITIZING AND SCREEN INTERFACE

Entry of graphical information into an FM/GIS system is achieved by digitizing hardware. A mouse can be used to digitize, since it provides a means of transferring a positional signal to the screen. Many digitizing activities can be performed with a standard mouse, and the user interfaces with the on-screen display. In addition, a mouse may have two or three buttons to enhance its functionality. The additional buttons can provide intelligence to the interaction, where the screen click may search for specific objects in the neighborhood of the click position. These functions are generally more important to CAD operators or programmers than to system users and require a higher level of sophistication in the user.

Drawing production activities where large-scale digitization is being conducted require a digitizing table or tablet. They provide the means for establishing accurate graphical representations of facility features. High-quality units are needed to meet mapping standard or engineering construction levels of accuracy. There are also desktop units for this purpose that can be used for smaller size drawings. Larger units are the size of conventional drawing boards and are designed to accommodate engineering size drawings. When selecting the appropriate sized unit for a particular application, it is better to have a slightly larger unit than may be needed than one that is too small. These devices have positional accuracy that is related to their initialization and physical design during a digitizing session and they also have commands that can be activated by clicking on special locations on the board. These are somewhat analogous to special function keys on a keyboard. Digitizing tables are not generally needed by most users of the ES&H system.

F. PLOTTERS AND PRINTERS

Plotters for GIS applications can range from small 8.5- × 11-in. desktop units to full engineering size units designed to print on roll paper or in sheets. Pen plotter technology has been overtaken by ink jet and electrostatic plotting. Pen plotting requires navigation around the page on a flat bed or drum plotter to produce all of the vectors in the drawing. Ink jet or electrostatic plotters perform repeated passes over the page and deposit the drawing in a raster image. Ink jet plotters operate in conventional office environments, but space conditioning to control humidity and temperature is generally advised for electrostatic plotters, which are affected by atmospheric conditions. The plot files required by raster plotters are larger than comparable pen plotting units, but they produce drawings of high quality, faster, and with fewer moving parts. These units are more costly, particularly color electrostatic plotters.

Plotters are mechanical equipment and, as such, they can be a weak link in the production process. Plotters fail much more frequently than computers; this is sometimes surprising to personnel who are introduced to plotting, even after significant experience with computer hardware. Mechanical maintenance activities associated with personal computers or workstations are not as extensive as they are for plotters, which require an effective preventive maintenance program to keep them fit and operable. This includes continuing efforts to assure that pens and ink cartridges are kept clean and that they are available in all desired colors. These supplies and paper often run out or fail during a production crisis for the uninitiated user of plotters.

Printers can be used for plotting purposes in many small applications. Drawing images can usually be converted to graphics formats that are acceptable to ink jet printers. This is effective for small jobs and is advisable as a preliminary viewing mechanism for large drawings.

II. SOFTWARE

As with any typical emerging technology, a large number of vendors have entered the CAD, GIS, and FM markets. Their offerings are attractive from the standpoint of graphical appeal, and it is often difficult to distinguish their differences. The importance of these differences will depend on the nature of the applications to which they are directed.

A. OBJECT- AND LAYER-BASED SYSTEMS/RASTER-BASED GIS DATA MODELS

Three general types of software approaches have been implemented for GIS applications. Layer-based systems grew out of automated mapping technology, in which overlay processing of discrete layers was needed to make decisions for broad geographical areas. Raster-based systems have been developed for resource management, such as interpreting data from satellite imaging systems. Object-based systems grew out of the need for automated engineering drawings where distinct components of the drawing had importance. Aspects of each of these approaches have been implemented for a variety of applications in the CAD/AM/FM/GIS arena.

Object-based GIS provides greater flexibility for the facility management applications of the ES&H requirements than does layer-based GIS. FM is a discipline requiring access to engineering and scientific characteristics of buildings, equipment, and locations, which are objects on engineering drawings. There is little requirement to process polygon layers in FM applications; however, object-based GIS drawings can be structured in layers to afford compatibility with such applications. Likewise, layer-based systems have been modified to attain object-like appearance. Users should understand the differences between these approaches and the impacts on performance. These impacts can occur in the creation of drawings, the access to objects, and the linkages between objects and databases.

B. SOFTWARE STRUCTURE

Software for GIS applications has evolved from minicomputer and workstation environments. As a consequence, it is available in the following general structures:

- Designed for workstations
- Designed for workstations or minicomputers/migrated to PCs
- Designed for PCs

In addition, the software may be derived from a CAD or from an AM genesis. The objective of GIS technology is to link the G (geography) with the I (information) to aid in spatial analysis. The extent to which this objective is achieved depends on the inherent design of the software.

Software designed for automated mapping (AM) applications is intended to process graphical objects and preserve their spatial relations in accordance with a specified projection scheme. There are little data associated with AM applications, other than the name of the geometric component and perhaps a few data elements describing its graphical appearance. The same is true with CAD applications, which are electronic graphical representations of what had been traditionally done on a drawing board. Attribute tables are used with these systems to store the small quantities of data elements, usually line styles, etc., with their associated graphical objects.

The use of attribute tables has been extended to handle more data types and is often used for linking objects to simple data tables. However, this approach is limited in its capacity to deal effectively with large numerical data tables and with relational data sets. A DBMS is needed for this purpose. Linking AM- or CAD-based systems to DBMS technology has become a principal item of concern to vendors in the GIS arena. These links are being retrofitted to conventional AM or CAD software in many instances to provide the user communities with preliminary capabilities for FM and GIS functionality. These retrofits have differing levels of effectiveness and success in achieving their objectives. Obviously, designed-in functionality for relational DBMS links to objects is more elegant and will perform better.

C. SELECTING SOFTWARE

The appropriateness of a given FM/GIS software tool to a specific need, particularly the FM applications for ES&H that are the subject of this writing, should be evaluated before it is selected for this purpose. The items of concern to the implementing organization include:

- Whether it is object-based
- How the data model fits the application
- The inherent DBMS
- Flexibility and efficiency of interface with organization's DBMSs
- Programming languages and macros for creating new functionality
- Applications to traditional FM functions
- ES&H functionality
- Operability in a GUI environment
- Hardware platforms supported by and compatibile with the organization's hardware
- Ability to interface with organization's existing CAD systems
- Engineering design functionality and features
- Drawing accuracies
- Performance at expected user levels
- Acquisition costs of hardware and software for initial user base

- Training and support costs
- Economies of scale for large organization-wide implementation
- Life cycle costs, including hardware, software, drawing and database loading, and integration
- Soundness of the structured evolution program

The relative priorities of these items should be established by the ES&H users and the MIS group. Alternative software platforms for the system can be evaluated by preparing comparative summaries based on vendor literature, demonstrations, and responses to information requests. This will reduce the vendors to a limited number who have met the requirements and from whom formal proposals can be requested.

III. SYSTEM USE

The ES&H system is intended to provide a standardized approach for accessing data and drawings, developing information products, and making the data and information products available to those who need them. It is expected that all users will not have the same needs. The specific requirements for a given organization will dictate how the system is installed within that organization.

A. ADMINISTRATION FUNCTIONS

The administration of the system will require typical data management functions that are common to any multiuser system. These are:

- Loading drawings and editing them
- Making drawings intelligent so they can interface with the DBMS
- Managing the DBMS table contents
- Managing user access
- Updating and archiving files
- Updating software
- Training
- Monitoring performance and user needs

ES&H staff in the plants or at corporate office locations may have specific responsibilities for discrete components of the system. Plant personnel may have responsibility for drawings or data associated with their facilities, or the corporate engineering department may have responsibility for all facility drawing. The corporate ES&H group may be responsible for selected tables in the DBMS that affect all plant locations. The appropriate administration procedures must be developed to fit the ES&H organization and will depend on the levels of involvement of other organizational elements. There are a variety of approaches for distributing the responsibilities for managing the system. Active involvement of ES&H staff in the process, working in conjunction with their MIS counterparts, is essential to a successful system.

B. ACCESS TO DATA AND INFORMATION PRODUCTS

Users of the ES&H system will need access to information at different levels and for different purposes. During an incident (accidental release, emergency, accident) top-level managers will want immediate information concerning the facility and the circumstances. In conjunction with a site investigation, ES&H managers may want to see the latest field sampling results or the impact of alternative cleanup levels on costs. These users will want different levels of access to the system than those who may have responsibility for a single plant or for a specific set of regulations. Furthermore, managers are likely to be at locations that are different from the facility of interest, and the database and drawings may be at yet another location. These circumstances must be considered in providing access to system functions.

1. VIEW-ONLY ACCESS

Many GIS systems include a viewing level of functionality that is appropriate to the type of access needed by managers. This *view-only* capability is designed to use available data and drawings to produce information graphics from them, but does not provide the capability that is available in the full system configuration for changing drawings permanently. This is an ideal approach for achieving lower implementation costs, because the associated license costs for these limited capabilities are significantly less (perhaps 75% lower) than for full system functionality. The system requirements phase of implementation should identify opportunities for applying view-only capabilities, which are somewhat analogous to *read-only* capabilities associated with text and database access. Furthermore, many other users who are not maintaining drawings and databases may be served effectively by the view-only version of the FM/GIS software.

2. RESPONSE PERFORMANCE

A second item of concern is the access speed for obtaining screen graphics. The issues here are similar to those identified earlier with regard to responsiveness of the system. If the view-only managers are co-located with the data and drawing source, LAN or direct connections to the FM/GIS software can be implemented to provide the same responsiveness as that is available to daily users of full system capability. If, on the other hand, users are located at distances that can only be served through telecommunications lines, the capacities of these lines becomes the critical factor in determining access speeds.

If a drawing has a size of 1 MB and it is transferred over a line served by a conventional data modem having 9.6 KB speed, it will take over 1.5 min to transfer the drawing $(1,000,000/[9,600 \times 60])$. Suppose the drawing, which represents site conditions for an incident or cleanup, is being managed at location A, is to be reviewed by a management team at location B, and is to be presented to a management team at location C. This can be achieved by developing hard-copy representations and sending them from A to B for review and comments, requiring 2 days of overnight mail prior to producing a final product. The 1.5-min transfer time could result in several reviews and final product preparation within a few hours, which is an appreciable time savings. Under these conditions, the 1.5-min drawing rate does not seem unacceptable. However, to the user who may be interested in zooming into selected areas of the drawing from a remote location, the 1.5-min penalty for each view may become tedious and unacceptable.

3. ALTERNATIVES FOR VIEWING GRAPHICS

It is important to develop appropriate requirements and related expectations when remote access to drawings is being contemplated. There are a number of options available for achieving this objective, and they each have related performance levels and costs. These will depend on where the drawings are located, where they need to be viewed, and the system configuration that is available. Examples of these alternatives are summarized below.

- Local network access — If the system resides on a PC or workstation or on a minicomputer designed for multiple users in a single geographic location, direct access using Ethernet can be provided to the system with rapid (few seconds) response. Access may be through a PC on which the system is installed or through graphics terminals or PC terminal emulation to a minicomputer. The computer may need the capacity to handle multiple users and an appropriate number of licenses to handle the expected number of concurrent users.
- Remote telecommunications access — The same approach as for local network access can be implemented with dial-up lines into the ES&H system computer. Performance will be much slower due to line speed limitations, but may be acceptable for limited or casual access needs. For a multiuser system, the number of incoming lines must be adequate to handle the expected number of concurrent dial-up users. Leased high-speed lines offer an alternative for improving performance and quality of use.

- Drawing file transfer — This approach calls for sending the completed drawing from the ES&H computer to the viewing computer. This can be achieved through wide area electronic mail or through a dial-up modem exchange using a commercial communications package. Diskettes could also be mailed as a slower alternative. The remote user will need a view-only license of the GIS software. The remote user does not directly interface with the ES&H computer where the file resides.
- Drawing conversion and transfer — Remote users may already have CAD capabilities that can be used to view the drawing. The drawing could be converted from the GIS format to the CAD format and then sent as with the above option. There would be no need to obtain additional licenses, but the conversion and transfer steps should be fully automated to make the process efficient.

In addition to these conventional methods, there are techniques for image file compression that can be effective. Commercial software is available to compress files by factors of about four or five through compressing a single line in the image and several following lines, and conveying only the differences. Characters are used to represent an image characteristic and the number of repetitions of the characteristic that follow in the image file. These characters add intelligence to the file and can reduce its size dramatically. Additional compression can be achieved by windowing on selected areas of the drawing and transferring only areas of interest rather than the full drawing. By decreasing the size of the file being transmitted, response times can be reduced significantly, say from the 1.5 min to, perhaps, less than 15 s. This requires the integration of the file compression software into the user interface. The extent to which this is feasible will depend on the system configuration, the drawing file format, and the software that is selected. In an ideally configured environment, the users should not be concerned with or aware of the compression/decompression process, which should be transparent to them in the same manner as modulation and demodulation are used to telecommunicate digital signals over analog networks. (Telephone lines are analog networks that are used for computer communications; the digital computer signal is transformed to an analog signal [like voice] for transmission, and then retransformed by a modem at the other end of the line.) Implementation of this approach can assist in providing direct access to drawings using terminals and dial-up lines from remote locations.

I. SYSTEM IMPLEMENTATION

The implementation of an ES&H system can be conducted through a traditional approach or through structured evolution.

A. TRADITIONAL APPROACH

Traditional approaches for system development have been defined to assure that the implementation is in accordance with standards and meets the organization's requirements. These are used by MIS organizations to develop new software applications. While they vary to some extent, the general approach includes the following general steps:

- Define the need and objectives of the system
- Develop the requirements for satisfying the need
- Develop a schedule and cost estimate for completion
- Design a system structure that fulfills the requirements
- Prepare a detailed design of system components and interfaces
- Develop a prototype of the system
- Test the system
- Implement the system

This approach is an effective method for creating a new software system. However, depending on the objectives and scope of the ES&H system, it can be a lengthy process. Many projects have difficulty with this process, which requires definitive products that can be used to transition the information developed from planning and design to the programming specialists who will implement the system. The time periods from the first to the last step can be several years. Regulatory, technology, market, and organizational changes can occur that confound the implementation. Application software contractors may be required to complete the implementation. The requirements may be difficult to document by MIS staff not having experience with the regulations. Cost estimates made by MIS and ES&H staff to budget the project may be lower than actual bids from contractors. Frustrations occur when these factors cause further delays to an already lengthy project. Intermediate products are exhaustive reports describing the steps rather than the tools that can be used to meet the user needs.

ES&H managers want short-term solutions, perhaps incremental, rather than long waits for ultimate systems that will produce all-encompassing capabilities. Staff want tools they can use to manage the ever-growing requirement for data collection and reporting. Disenchantment with the traditional approach often results in the selection of a point solution that does not fit into an overall plan but addresses an immediate need. This approach is not likely to integrate with the other point solutions adopted earlier for the same reasons.

Modern software tools facilitate software development, providing means for creating these incremental steps quickly and for demonstrating continuing progress to managers. Because of the diversity of requirements and knowledge levels, it is important to demonstrate progress with software rather than with study results and design reports. Readers can interpret these documents differently. For example:

- An ES&H team develops a requirements document and submits it to an MIS group for implementation
- The MIS group enhances the document with information technology requirements and creates a request for proposals
- Contractors are solicited to submit proposals
- Proposals of varying types, scope, and cost are submitted, all in response to the same document

This is because of interpretation. The same is true of the design documents that are produced during the implementation of the system, except that there is one document and varying interpretations by the reviewers and users. Users would prefer to see prototypes that have the look and feel they want as well as simulated performance or actual functional performance in limited capacities. Furthermore, because of the need to use existing systems and technology in the ES&H system, they are better served by a structured evolution that uses many of the principles of standard software development, but acknowledges the underlying need to integrate the existing tools that are available to them.

B. STRUCTURED EVOLUTION

Effective ES&H information management in a large organization requires the active participation of many organizational components involved in collecting, storing, maintaining, creating, and using data and information products. Coordination of these participants must take on a level of importance that is appropriate to the objectives and proportional to benefits to be derived from the system.

The current status of the organization may be one of loose coordination, if any, among the players. They may not even realize that they are players. To be players, there must be a game, and the players must be interested in the game. Only the top management of an organization can effectively focus this group into carrying out their roles. This may require substantive reengineering of processes and procedures, with a commensurate affect on roles and responsibilities. Interdepartmental projects and systems can lose momentum quickly if the coordination process is not given attention. Middle managers and technical staff can become frustrated over competing policies unless a top management direction is established and progress is monitored.

Building an integrated system requires a number of important elements for success. These include:

- A vision
- Responsibilities
- Schedule and milestones
- Measures of performance

1. VISION

Vision is a trait that is essential to overcoming organizational inertia. There are always reasons why something cannot be achieved, but it takes vision to define what needs to be done and to place a priority on its implementation. FM/GIS is an emerging technology that will change how organizations integrate their information. Those organizations who can create a vision in this area and convince themselves to proceed with its implementation can reap significant benefits as their employees learn how to take advantage of the power of this technology.

2. RESPONSIBILITIES

It is evident that the approach advocated for ES&H system in this book requires participants from several organizations. The ES&H organization must take the lead in implementing its vision, and someone must be in charge of the overall program, which will affect other organizational entities. Empowering a manager to implement a system of this type is essential to its success. Someone must take charge and have the authority to move forward with the project steps. Because the project is inherently one of integration, it will be necessary to have several key team members who drive the system implementation. They must be knowledgeable in the ES&H need, the organizational direction with regard to its information resources and infrastructure, and the FM/GIS technologies that are being implemented. The success of the project depends on a solid team that can identify needs, develop plans, manage implementation, and communicate status.

3. SCHEDULE AND MILESTONES

There should be definitive objectives related to specific milestone dates associated with the evolution of the system and a schedule for its implementation. Milestones can be items such as completion of a pilot project or loading drawings for a given facility or process diagrams for a business unit. A set of milestones should relate to one round of updating and maintenance of the facility drawings and associated land base data to convey the need for continuing maintenance of the system. Each milestone should relate to the implementation plan and provide managers with visible measures of system status and availability. The schedule can be portrayed by a line graph illustrating planned quantities and actual levels for items such as database size, number of drawings installed, number of user hours, and implementation costs. These can be illustrated at the business unit or corporate level. Managers need continuing awareness of progress and simple measures to illustrate status. The structured evolution process can be fostered by continuing attainment of milestones within planned budget.

4. MEASURES OF PERFORMANCE

In addition to implementation goals, there should be performance goals associated with the implementation. The project can only be successful if there are continuing benefits in evidence. The success can be measurable through quantitative techniques such as cost reductions, productivity improvements, or reduced durations to achieve compliance submittals or other information products. Quantitative performance measures can be illustrated in the same manner as schedule goals. Qualitative information can also be used to assess status and identify direction for improvement. User surveys can elicit feelings about system performance in achieving job requirements or its flexibility in responding to unexpected queries. If continuing benefits are derived, the structured evolution can proceed.

REFERENCES

1. Business Geographics, *National Association of Realtors Adopts GIS,* Marcie Wallis, May/June 1993, Vol. 1, No. 3, published by GIS World, Inc.

2. *Business Geographics,* September/October 1993, Vol. 1, No. 5, published by GIS World, Inc.

3. Business Week, *Consultant, Reengineer Thyself,* Louis Therrien, April 12, 1993, pp. 86, 88, published by McGraw-Hill.

4. *Earth Observation Magazine,* July/August 1993, published by EOM, Inc.

5. *Environment Today,* January 1993, Vol. 4 No. 1, p. 41, published by Enterprise Communications, Inc.

6. Geo Info Systems, *GIS Approaches to Resolve Resource Conflicts in the Himalayas,* Hans Schreier and Sandra Brown, October 1992, Vol. 2, No. 8, pp. 52-58. published by Aster Publishing Co.

7. Geo Info Systems, *Managing Oil Spills,* Michael Garrett and Gary A. Jeffress, January 1993, Vol. 3, No. 1, pp. 29-35, published by Aster Publishing Co.

8. Geo Info Systems, *Analyzing the Cumulative Effects of Forest Practices: Where Do We Start?,* Kass Green, et. al., pp. 31-41, February 1993, Vol. 3, No. 2, published by Aster Publishing Co.

9. Geo Info Systems, *Gulf War Legacy — Using Remote Sensing to Assess Habitat in the Saudi Arabian Gulf Before the Gulf War Oil Spill,* Sumi Narumaiani, et. al., pp. 33-41, June 1993, Vol. 3, No. 6, published by Aster Publishing Co.

10. GIS World, *GIS Improves Visualization, Evaluation Capabilities in Superfund Cleanup,* Joseph M. Gracia and Lous G. Hecht, Jr., February 1993, Vol. 6, No. 2, pp. 37-41, published by GIS World, Inc.

11. Gis World, *Digital Orthophotography Bolsters GIS Base for Wetlands Project,* Bryan J. Logan, June 1993, Special Issue, pp. 58-60, published by GIS World, Inc.

12. GIS World, August 1993, Vol. 6, No. 8, published by GIS World, Inc.

13. Gis World, *Follow a Seven-Step Path to GIS Nirvana,* Andrew W. Flagg, September 1993, Vol. 6, No. 9, pp. 46,47, published by GIS World, Inc.

14. Government Technology, *GIS on the Comeback Trail,* Tod Newcombe, July 1993, Vol. 6, No. 7, Published by GT Publications, Inc.

15. Hazmat World, *Linking Databases to Plant Drawings Saves Time and Money in Process Hazard Analysis,* Cindy Lancaster, July 1993, p. 76.

16. Public Works, *Opportunities for Funding a GIS,* Steve Statler, February 1993, Vol. 124, No. 2, pp. 34-37, published by Public Works Journal Corp.

17. Antenucci, John C. and Kay Brown, Peter L. Croswell, Michael J. Kevany with H. Archer, 1991, *Geographic Information Systems,* (New York: Van Nostrand Reinhold Company)

18. Michael, Gene Y., 1991, *Environmental Data Bases,* (Chelsea, MI: Lewis Publishers, Inc.)

19. Douglas, W. J., Soby, M. T., Haasbeek, J., and Klaber, K. Z., *Air Emissions Data Management to Meet Clean Air Act Requirements,* American Chemical Society, 144th Fall Int. Tech. Meeting, Orlando, Florida, October 17, 1993.

20. Maitin, I., and Klaber, K. Z., *Geographic Information Systems as a Tool for Integrated Air Dispersion Modeling,* Proc. GIS/LIS '93 Annu. Conf. and Exposition, Minneapolis, MN, November 2 to 4, 1993.
21. Michael, G. Y., *Environmental Data Bases,* Lewis Publishers, Chelsea, MI, 1991.
22. Radian Corporation, *Air Chief CD-ROM User's Manual,* EPA45/4 92-015, U.S. Environmental Protection Agency, Research Triangle Park, NC, May 1992.
23. Hamilton, S. D., New GIS Tools, in *Digital Orthophotography and Integrated Raster and Vector Data Types, An Overview,* ADR Associates, Pennsaucken, NJ, 1990.
24. Nale, D. K., *Digital Orthophotography, What It Is and What It Isn't,* ADR Associates, Pennsauchken, NJ, 1993.

INDEX

A

Access, 110–112
Accident risk, 3
Administration functions, 109–110
Advanced Visualization System
    (AVS), 48, 70
Aerial photos, 4, 9
Aerial surveys
  contours produced by, 22
  data from, 44–45
  drawing specifications from, 17–18
  drawings and maps from, 17–20
  historic map files with, 37
  quality assurance in, 18–19
Air dispersion models, 50
Air emissions, 3, 11
Analytical method, 40
ASCII files, 65
Attribute tables, 108
Automated digitization, 24–25
Automated mapping (AM) software,
    107–108
Automated mapping/facility manage-
    ment system (AM/FM), 7
AVS, see Advanced Visualization
    System (AVS)

B

Batch loading, 91
Bit map, 17
Bore hole logs, 65

C

CAD, see Computer-aided design
CD-ROM systems, 33, 50, 95
CD storage media, 33

Chemical contamination levels, 3
Chemical inventory, 30
Chemical inventory table, 15
Chemical laboratory analyses, 38–
    41
Clean Air Act Amendments, Title
    V, 30
Clearinghouse for Inventories and
    Emission Factors, 50
Coding forms, 43
Commercial planning, 11–12
Compliance management systems,
    29, 30
  implementing of, 31–32
  least-cost strategies in, 3–4
  quantifying requirements in, 30–
    31
  regulations, documents, and
    electronic mail in, 32–35
Compliance reports, 54
Compliance software integration,
    60–61
Comprehensive Environmental
    Response, Compensation and
    Liability Act (CERCLA), 27,
    35
Computer-aided design
  applications of, 107–108
  drawings, 17, 22–23
    accuracy of, 13–14
    conversion in, 22–23
    in facility management and GIS,
    9
    in infrastructure management, 8–
    9
    integration of, 26–27
  files in, 90
  format, base map in, 58
  generic systems of, 45–46

Computer-aided software engineering (CASE), 80, 87
Computers, see also Mainframes; Personal computers; Workstations
  databases for, 12; see also Databases
  digitizing and screen interface in, 105–106
  hardware of, 47
    in facility management, 56
    in system management, 101–106
  memory and disk storage in, 104
  monitors and displays in, 105
  networks, 80
  open architectures and distributed environments in, 80
  portable, 42
  in site visualization, 46–48
  software for, see Software
Computing infrastructures, 79
Concentration values, 40
ConSolve, 48
ConSolve Site Planner, 70–73
Contamination
  measurement of, 3
  plume visualization, 69–70
Contouring, 40, 62
Contouring software, 65
Contours, 22
Controlled access, 20
Coordinates, 15
Corporate image, 6
Corporate vision, 114, 115
Corridor planning, 12
Cost account codes, 63
Cost models, remediation, 49–50
Cost sharing, 10
Costs
  data on, 63
  ES&H, 5–6
  of integration and implementation, 67

Cross section visualization, 70, 74–75
Customized indexing, 96

D

Data
  access to, 110–112
  aerial and land survey, 44–45
  consistency of, 4–5
  entry of, 43–44
  extraction of, 60
  in facilities and infrastructure management, 9
  field, 42–44
  management systems for, 29
  types of, 36
Data dictionary, 88
Data field, 88
Data model
  elements of, 88–94
  information products of, 94–95
  purpose of, 87
  for system integration, 87–95
Data tables, 88
Database management systems, 7, 43, 60, 108
  hierarchical, 60–61
  for waste site investigation, 35–36
Database tables, 15, 36
  drill-down through, 75–77
  extraction from, 60
Databases, 65
  access to, 59–60
  elements of structure of, 88–94
  in facilities management, 12
  field data in, 42–44
  laboratory analysis in, 40
  linkage of, 81
  management of, 29–52
DBMS, see Database management systems
Decentralized strategy, 3–4
Decision criteria, 68

Defense Department, reporting requirements, 2–3
Design/implementation approach, 90
Detection limits, 40
Digital elevation model (DEM), 18, 21, 22, 44
Digital orthophotography, 19–20
Digitization
  automated, 38
  large-scale, 105–106
Digitized drawings, 13–14, 23–25
Digitized map files, 50
Digitizing tables, 106
Direct access, 60
Discharge database, 88–90
Disk storage, 104
Displays, 105
Distributed computing environments, 80
Document management systems, 96
Documentation, 67
Documents
  in facilities and infrastructure management, 9
  storage of, 33–34
Drawing file transfer, 112
Drawings
  access to, 20, 59
  accuracy of, 13–14
  from aerial surveys, 17–20
  CAD, 22–23
  changing features for, 63–64
  conversion of, 22–26, 112
  definition of, 13
  in facilities and infrastructure management, 9
  hard copy, 23–25
  information types in, 14
  integration of, 26–27
  intelligent, 13–28
  layers (phases) of, 14
  manipulation of, 59
  project coordination in, 20–22
  schematic, 25–26
  specifications of, 17–18
  storage of, 34
  structure and organization of, 14–17
  transfer of, 112
Dynamic visualization, 55

E

Earthvision, 48, 69
EDGE (Environmental Data Graphics from ERM), 48, 55, 63
  customization in, 65
  information access through, 67–68
  modules in, 60–61
  remediation design in, 64
  structure of, 56–58
Electronic data loggers, 42
Electronic document management system (EDMS), 34
Electronic file transfer requirements, 42
Electronic linkages, 15
Electronic mail, 34–35
Electronic regulations, 96–97
Electrostatic plotters, 106
Elevation contours, 21–22
Emergency information requests, 54
Emergency response, 9
Emergency services planning, 10
Emissions database, 88–90
Energy Department reporting requirements, 2–3
ENFLEX DATA, 60
Engineering drawings
  hard-copy, 23–25
  integration of, 26–27
Engineering feasibility studies, 49–50
Engineering workstations, 102–103
  memory and disk storage in, 104
  in site visualization, 46–48
Entities, 87

Entity diagrams, 88
Environmental facility management
  decentralized approaches in, 3–4
  graphics for, 73–77
  information needs for, 1–6
Environmental incidents, 9
Environmental information manage-
    ment system (EIMS), 32
Environmental integration, total, 56
Environmental planning, 11
Environmental site investigations, 4
Environmental Software Directory,
    32
Environmental studies, 11
EPA Air Chief, 50
EPA requirements, 36
ES&H system
  data for, 4–6
  database management for, 29–52
  facility management systems for,
    53–77
  implementation of, 112–116
  information needs of, 86
  integration of drawings in, 26–27
  management of, 3–4
  remediation program in, 51–52
  structured evolution of, 82–83
Evaluation criteria, 32, 66–67
Evaluation process, 66–67
Excavation volume/shape, 49, 64
Executive information systems
    (EIS), 67, 77
  approach of, 67–68
  for site remediation and
    prioritization, 91–94

F

Facets, 14
Facility database table, 58–59
Facility-level information, 58–59
Facility management systems, 7–12
  applications of, 9–10
  approach for, 53–77

geographic information system
    and, 7–8
  in government and industry, 10–12
  intelligent drawings for, 13–28
  prototype for, 55–67
Feasibility studies, engineering, 49–
    50
Federal regulations, 1–2, 95
Fence diagrams, 65
Field data, 42–44
Field mobile units, 42
Fields, 88
File compression, 112
File transfer, 65, 112
Floor plans, 15
Flow characteristics, 49
FM/GIS systems
  intelligent drawings for, 15–28
  map files for, 27–28
  object orientation of, 15
FMC Ground Systems, 48
FMC visualization system, 69
FOCUS DBMS, 60
Forestry management, 11
FoxBase database, 48
FoxPro database, 48

G

Gantt charts, 51
GDS Site Modeler utility, 62
GDS Solid Modeler utility, 62
Geo-Graphical User Interface, 56–
    58, 60, 76
Geographic information system
    (GIS), 7–8
  applications of, 9–10
  early attempts at, 82
  generic, 45–46
  implementation of, 82
  software, 20, 40
  technology in information visual-
    ization, 54–55
GEOKIT, 70

Geological data, 42–44
Geological data interpretation, 70
Geological Survey, U.S., 27
  digitized map files of, 50
Geologically focused software, 65
Geophysical logs, 4
Geostatistical methods, 21–22
Geotechnical conditions, 49
gINT, 70
GIS/Key, 48, 70
Government
  facilities management in, 10–11
  management and planning in, 10
Graphical drawing files, 55, 104
Graphical drill-down, 75–77
Graphical representations, 15–17,
    see also Drawings; Maps
Graphical user interface (GUI), 56–
  58
Graphics
  alternative viewing approaches to,
    111–112
  cost and schedule, 63
  environmental facility manage-
    ment, 73–77
  high-speed, 103
Green movement, 5–6
Gridding, 21
Gridding algorithm, 21–22
Ground water
  elevation tables, 15
  elevations of, 4
  models, 50
Groupware, 34–35

H

Habitat characterization, 11
Hard-copy engineering drawings,
  23–25
Hard-copy representations, 55
Hazardous chemicals records, 3
Hazardous Operations (HAZOP)
  analysis, 11

Hazardous waste site characteriza-
    tion, 35–45
Health planning, 11
Hierarchical format, 88, 89, 91
Historical data sets, 36–37
Historical map files, 37–38
Hits, 40
Human resources integration, 84–85

I

I/O capacity, 104
Image access, 62–63
Implementation
  of ES&H system, 112–116
  strategy for pilot project, 98–99
Implementing systems, 31–32
Industrial facilities representations,
  8
Industrial source complex model, 65
Industry, facilities management in,
    10–12
Information
  exchange of, 84–85
  executive systems of, 67–68
  facility-level, 58–59
  integration of, 86
  intuitive access to, 81
  management for waste site
    investigation, 35–45
  needs, 1–6
  regulatory, 95–99
  representations in facilities and
    infrastructure management, 9
  resource integration, 81
  resources for waste sites, 45
  technologies, 54, 79–80
Information flow, 84
  resistance to, 86
Information products, 46, 53–54,
    94–95
  access to, 110–112
  example of, 68–77
  types of, 54

visualization of, 54–55
Infrastructure management, 8–9
Initial conditions, 79
Ink jet printers, 106
Integrated regulatory information,
      96–99
Integrated system, 114–116
Interdepartmental cooperation, 114
International regulation require-
      ments, 1–2

K

Key fields, 88
Knowledge, 86

L

Laboratory analytical data, 38–41
Laboratory data structure, 90
Laboratory identity, 40
Laboratory information management
      system (LIMS), 39
Land surveyor, 21
Land surveys, 4, 44–45
Layer-based systems, 107
Local area network (LAN), 110–111
   interface in, 65
   technology of, 34
Local network access, 111
Local regulations, 96
Logical queries, 63–64
Loose coupling-file transfer, 65

M

Mainframe computers, 101, 103
Management
   information needs of, 1–6
   reports of, 54
Managers
   ES&H, 5–6

information needs of, 5–6
Map files
   digitized, 50
   historical, 37–38
   importing from public sources, 27–
      28
Mapping software, 40
Maps, see also Drawings; Graphical
      representations
   accuracy of, 13–14
   base, 58, 59
   definition of, 13
   in facilities and infrastructure
      management, 9
   historical, 37–38
   quad, 69
   schematic, 25–26
   specifications of, 17–18
   three-dimensional, 21–22
   two-dimensional base, 21
Market research, 11
Material safety data sheets (MSDS),
      30
   storage of, 33
Mathematical transformation
      software, 28
Memory, 104
Menus, 58, 60, 63–64
MGE system, 69–70
Milestones, 115–116
Minicomputers, 102–103
MIS department, 113–114
   in facility management, 53
   information technologies and, 80
   staff of, 31
Mobile field analytical services, 42
Modeling data, 4
Modem, 112
Monitors, 105
Motion, direction of, 79
Multimedia systems, 80
Multiprocessing, 54

Multiuser FM/GIS applications, 103
Multiuser network, 98–99
Municipal infrastructures, 10

N

Natural resource dynamics, 11
Numerical surface software, 65

O

Object-based programming, 80
Object-based systems, 15, 107, 108
Object intelligence, 59–60
Objects, 14, 17
Oil spills
  emergency responses to, 9
  impact of, 11
Olympic games planning, 10
Open architectures, 80
Oracle data table, 76
Organizational integration, 80, 85
Organizational planning, 10
Organizational policies, 5–6
Orthophotography, 19–20
OSHA requirements
  chemical inventory, 30
  for hazardous operations, 11

P

Parent-child hierarchy, 88–89, 91
Pathworks network, 65
PCB equipment inventorying, 30
Performance measures, 116
Personal computers
  memory and disk storage in, 104
  in site visualization, 46–48
  in system management, 101–103
  technology of, 56
  in visualization systems, 46–48,
    70–73

Photogrammetric compilation, 20
Photogrammetric measuring
    techniques, 21, 22
Photogrammetrist, 21
Photogrammetry, 17
Photographic images, 15
Photographs
  access to, 62–63
  from aerial surveys, 17–20
  in facilities and infrastructure
    management, 9
  screen displays of, 55
Pilot projects, 97–99
Piping diagram, 25
Pixel clouding, 24
Planimetrics, 17, 18
Planning
  commercial, 11–12
  for environmental, safety and
    health needs, 11
  government, 10
  reports, 54
  resource, 11
Plant condition visualization, 55
Plant personnel reporting, 2–5
Plotters, 106
Pop-up menus, 58
Population planning, 11
Prince William Sound oil spill, 9
Printers, 106
Prioritization
  criteria for, 68
  EIS for, 91–94
Process control, 66
Process drawing objects, 59
Process-flow diagrams, 25, 90
Process hazard analysis, 11
Process Line Performance Table,
    88–90
Program execution accessing, 60
Project coordination, 20–22
Prototype system, 55, 97–98

functions of, 58–64
software integration of, 64–67
structure of, 56–58
Public agencies, 27–28
Public risk, 3

Q

Qualitative information, 116
Quality assurance, 83–84
  in aerial drawings, 18–19
  for field data, 43
  in laboratory analysis, 38–39
Quality control sample, 40
Quality data, 4
Quantitative measurements, 3

R

Rapid prototyping, 97
Raster-based GIS data models, 107
Raster image, 17, 19–20
Read-only access, 33, 110
Real estate planning, 12
Red, Green, Blue (RGB) output,
  105
Re-engineering philosophy, 84
Regional planning, 10
Regulations
  access to, 32–33
  electronic access to, 29
  federal and state, 95
  integration issues in, 96–99
  local and special, 96
Regulatory information integration,
  95–99
Regulatory requirements, 1, 30
Relational database architecture, 80,
  91
Relational database management
  system (RDBMS), 66, 88
Remedial design, 51, 64
Remediation, 48–52
  construction phase of, 51–52

cost models of, 49–50
design of, 51, 64
EIS for, 91–94
Remote telecommunications access,
  111
Reporting requirements, 2–5
  data consistency in, 4–5
  organizational approaches to, 3–4
Requirements, quantifying, 30–31
Resource Conservation and Recovery Act (RCRA), 30
Resource planning, 11
Response performance, 110–111
Responsibilities, 115
Right to know, 3
Roof drawing, 25

S

Safety planning, 11
Sample location code, 40
Samples
  collection and analysis of, 38
  identity of, 39
Sampling locations, 15
Sampling program, 42
Scanned image, 24
Schedule, 115–116
Schedule data, 63
Schematic drawings, 25–26
Screen interface, 105–106
Site investigations
  database for, 91–92
  information management for, 35–
    45
  laboratory analyses in, 38–41
Site Planner, 48
Site remediation, 64, 91–94
Sites
  data on, 15
  maps of, 4; see also Maps
  remediation of, 48–52, 64, 91–94
  visualization of, 45–48
Siting/corridor planning, 12

Software
  compliance, 60–61
  in environmental management, 3, 32
  in facility management, 56
  selection of, 108–109
  specialized, integrating, 64–67
  structure of, 107–108
  in system management, 106–109
  tools of, 48, 113–114
  vendors of, 106–107
Software visualization tools, 48
Soil quality data, 4
Solid Waste Disposal Act (SWDA), 30
Spatial data modeling, 61–62
Specification development, 40
Spills, 3, 9
Spreadsheets, 36, 65
State regulations, 95
Statistical software, 65
Stratifact, 48
Stratifact geological analysis software, 70, 74
Stratigraphic column, 44
Stratigraphy, 4
Structured evolution, 80, 82–83, 114–116
Subsurface viewing, 69
Superfund Amendment and Reauthorization Act (SARA), 30, 35
Superfund legislation, 30, 35, 46
Surveyors, 44–45
System management, 101
  administration functions in, 109–110
  computer hardware in, 101–106
  data and information product access in, 110–112
  implementation of, 112–116
  software for, 106–109
Systems integration, 79
  data model in, 87–95
  factors affecting, 79–84
  information for, 86
  methodologies of, 80
  parties involved in, 84–85
  regulatory information in, 95–99
Systems integrators, 81

T

Tables, 88, see also Database tables
  entities of, 88–90
  parent/child, 88–89, 91
Tabular lists, 55
Tax management, 10
Three-dimensional elevation models, 21–22
Three-dimensional modeling, 21–22, 103
Topography
  modeling of, 21–22, 61, 103
  two-dimensional, 21, 44
Total quality management, 83–84
Toxic substances, 3
Toxic Substances Control Act (TSCA), 30
Traditional system development, 113–114
Transportation decision support, 10
Triangulated irregular network (TIN) model, 62
Triangulation techniques, 21–22
Two-dimensional topography, 21, 44

U

Urban and Regional Information Systems Association (URISA), 9
Urban development planning, 11

V

Value qualifiers, 40

VAX workstations, 47
Vector processing, 103
Vendor Information System for
    Innovative Treatment Tech-
    nologies (VISITT), EPA,
    49
Video displays, 105
Video Graphics Array (VGA),
    105
View-only access, 110
Virtual site, 70
Vision, 114, 115
Visualization, 45–46
  of contamination plumes, 69–70
  high- and low-end approaches in,
    46–48
  of information products, 54–55
  software tools for, 48, 65
  spatial data, 61–62
Visualization systems, 70–73
  features of, 71
  high-end, 69–70
  intermediate, 70–72
Voice recognition, 42
Volumetric computations, 21,
    62

W

Waste sites
  characterization of, 35–45
  information management for
    investigations of, 35–45
  information resources for, 45
  maps of, 55
  sampling locations of, 15
Wastewater discharge, 3
Wastewater management, 11
Water quality, 4, 11
Wide area network (WAN)
  interface, 65
  technology of, 34
Wildlife management, 11
Windows environment, 54, 57
Work breakdown structure (WBS),
    63
Worker exposure levels, 3
Workstations, 80
  contamination plume visualization
    using, 69–70
  memory and disk storage in, 104
  PC and minicomputer, 102–103
  technology of, 56